INVITATIONS
TO
SCIENCE INQUIRY

SUPPLEMENT
TO FIRST & SECOND EDITION

BY

Tik L. Liem

SCIENCE EDUCATION CONSULTANT

SCIENCE INQUIRY ENTERPRISES
14358 VILLAGE VIEW LANE
CHINO HILLS, CA 91709

This Supplement to the First and Second Edition has been published by:
SCIENCE INQUIRY ENTERPRISES
14358 VillageView Lane
Chino Hills, California 91709

Copyright ©1991 by **Tik L. Liem**
All rights reserved.
Drawings and sketches by the author and Dave Kristedja.

Permission in writing must be obtained from the publisher before any part of this work may be reproduced or transmitted in any form or by any means, electronic or mechanical, including photocopying and recording, or by any information storage or retrieval system.

ISBN: 1-878106-01-5

Printed in the United States of America

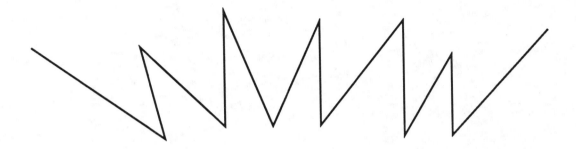

TO THOSE TEACHERS
WHO WANT TO PUT SOME *Excitement*
IN THEIR TEACHING

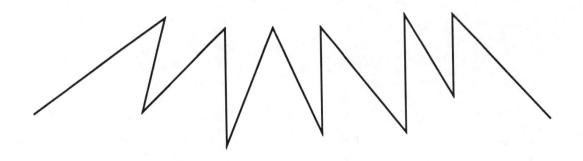

PREFACE

This supplement to the first and second edition has been compiled as a response to the many requests from those who already own the first or the second edition. It consists of the difference between the first and the second edition of "INVITATIONS TO SCIENCE INQUIRY", plus an additional **50 discrepant events** on top of the second edition. The difference between the first and second edition is an **expanded introduction** plus about **100 more discrepant events**.

Naturally it cannot contain exactly all differences between the two editions, because with every new printing or edition, I am attempting to revise and improve the older material. Obviously revisions of the discrepant events in the first edition cannot be included in this supplement. But since the **introduction** has been very much expanded, and since it forms the basis and philosophy of my inquiry approach, I have included the whole introduction as it appears in the second edition, with some improvements.

The main reason for compiling this supplement is to give everybody, those that own the first and those that own the second edition, even more **discrepant events** to enjoy and get excited about.

The format for each of the discrepant events on each of the pages has not been changed, only no numbering has been assigned to the new ones. The sequence of chapters has been maintained and for each chapter some new discrepant events has been added. A separate index has been compiled in this supplement for your convenience.

Finally, may I express again my gratitude to all the dedicated science teachers - mostly high school teachers and colleagues - who shared with me many of the ideas in this supplement.

Tik L. Liem
14358 Village View Lane
Chino Hills, California 91709 USA
(714)590-4618

CONTENTS

INTRODUCTION 1

    1) SCIENCE TEACHER'S CHARACTERISTICS 2
        A) ENTHUSIASM 2
        B) A HIGH SELF-ESTEEM 3
            I. DEVELOP THE ATTITUDE OF "I'M ACTIVATED". 3
            II. DEVELOP THE ATTITUDE OF "YOU ARE IMPORTANT". 4
            III. DEVELOP THE ATTITUDE OF "I CARE FOR YOU". 5
        C) CREATIVITY 6
        D) RESPONSIBILITY 7
        E) A SENSE OF HUMOR 7
        F) COMMUNICATION 8
            SUMMARY: TEACHER'S CHARACTERISTICS 9

    2) THE NATURE OF SCIENCE AND THE SCIENTIST 10
        A) THE NATURE OF SCIENCE 10
        B) THE NATURE OF THE SCIENTIST 11

    3) THE LEARNING PROCESS 13
        A) AROUSE CURIOSITY 13
        B) USE SIMPLE MATERIALS 13
        C) USE ALL GATEWAYS 14
        D) PROVIDE CONTEXT 14
        E) DO IT WITH JOY AND ENTHUSIASM 14
            SUMMARY: CONDITIONS FOR LEARNING 15

    4) THE DISCREPANT EVENT 16
        A) PSYCHOLOGICAL BACKGROUND 16
        B) WHAT IS A DISCREPANT EVENT? 16
        C) HOW TO USE DISCREPANT EVENTS? 17
        D) THE EMPIRICAL EVIDENCE 18

    THE FORMAT 19

    REFERENCES 20

ADDITIONAL DISCREPANT EVENTS 23

AIR
    THE WEIGHTLESS WATER 24
        EXERTS PRESSURE 24
    THE BALLOON IN THE JAR 25
        EXERTS PRESSURE; EXPANDS WHEN HEATED 25

THE CRUSHING POP CAN. 26
    EXERTING PRESSURE 26
TURN A LITTLE WATER INTO A LOT OF LEMONADE 27
    TECHNOLOGICAL APPLICATION; TRANSFERABILITY OF PRESSURE 27
MAKE AN AIR BULLET SHOOTER 28
    AERODYNAMICS 28
THE SMOKE RING RACE 29
    AERODYNAMICS 29

FLOWING AIR
THE LEAPING EGG 30
    WIND ENERGY 30
    BERNOULLI'S PRINCIPLE 30
LIFT A MOTH BALL WITH STREAMING WATER 31
    FLOWING FLUID; BERNOULLI'S PRINCIPLE 31
THE SELF-PRIMING SIPHON. 32
    FLOWING FLUID; APPLICATIONS 32
THE SELF-DIRECTING CARDS 33
    RESISTANCE 33

WEATHER
MAKE A FIRE CYCLONE 34
    CYCLONE 34

CHARACTERISTICS OF MATTER
THE COOLING RUBBER BAND 35
    MOLECULAR NATURE 35
GET THE THREE STATES OF MATTER OUT OF WOOD! 36
    STATES OF MATTER. 36
MAKE MILK FROM WATER AND OIL. 37
    COLLOIDS; STATES OF MATTER 37
THE INVISIBLE STEAM 38
    STEAM & MOISTURE; STATES OF MATTER 38
IS THE CUP REALLY FULL? 39
    MOLECULAR SPACING 39
THE MAGNETIC FINGER. 40
    SINKING & FLOATING; COMPRESSABILITY 40
THE CLEARING SPOT. 41
    DENSITY; IMMISCIBILITY 41
THE EXACT PAPER CIRCLE 42
    KINDLING POINT 42
THE FUNNY TOOTHPICKS 43
    ABSORPTION; SURFACE TENSION. 43
HOW MANY PENNIES CAN GO IN ? 44
    SURFACE TENSION 44
THE STRONG SOAP FILM 45
    CAPILARITY; COHESION & ADHESION; SURFACE TENSION 45

THE FLOATING OIL SPHERE 46
    DENSITY; SURFACE TENSION 46
WHICH WILL FLOAT WHERE? 47
    DENSITY 47
THE UNLEVEL COMMUNICATING TUBE 48
    DENSITY 48
THE COMMUNICATING CYLINDERS 49
    DENSITY; WATER PRESSURE 49
THE PLASTIC TUBING WATER LEVEL 50
    DENSITY; BUOYANCY 51
MAKE YOUR OWN LETTER SCALE! 51
    DENSITY; BUOYANCY 51
THE WATER CANDLE? 52
    BUOYANCY; MELTING POINT 52
THE FLOATING SOAP BUBBLE 53
    DENSITY OF GASES 53
THE DRINKING BIRD 54
    HEAT OF VAPORIZATION; VAPOR PRESSURE 54

CHEMISTRY
WALK THROUGH A HOLE IN ORDINARY NOTEBOOK PAPER 55
    PHYSICAL CHANGES 55
DRAW WITH FIRE 56
    OXIDATION REACTION 56
TURN A RED ROSE INTO A WHLTE ONE. 57
    REDUCTION REACTION 57
TURN WATER INTO WINE, MILK, AND BEER! 58
    PRECIPITATES; GAS PRODUCTION 58
THE EXPLODING BALLOONS 59
    COMBUSTION; PHOTOCHEMICAL REACTION 59
THE AMMONIA FOUNTAIN 60
    SOLUBILITY 60
BURN PAPER WITH ICE 61
    SPONTANEOUS COMBUSTION 61
THE POTASSIUM CHLORATE BOMBS 62
    SPONTANEOUS COMBUSTION; GAS PRODUCTION 62
BURN A PIECE OF METAL IN WATER 63
    SPONTANEOUS COMBUSTION 63
THE DELAYED EXPLOSION 64
    COMBUSTION; EXOTHERMIC REACTION 64
THE STRAW FLAME THROWER 65
    COMBUSTION; KINDLING POINT 65
THE HUMAN FLAME THROWER 66
    FLAME AND BURNING 66
THE MAGIC CANDLE 67
    FLAME AND BURNING 67

THE GLOWING ALUMINUM	68
REACTION ENERGY	68
TOUCH A CRACKER!	69
REACTION ENERGY	69
THE STICKY BOARD	70
ENDOTHERMIC REACTION	70
SHAKING THE BLUES	71
REVERSIBLE REACTION	71
THE COLOR ABSORBING BACON	72
ORGANIC CHEMISTRY; UNSATURATED BONDS	72
MAKE A NYLON THREAD OUT OF TWO LIQUIDS	73
ORGANIC CHEMISTRY; HIGH POLYMERS	73
THE STICKY MATCHES	74
TECHNOLOGICAL APPLICATION	74
MAGIC FLASH PAPER	75
COMPLETE COMBUSTION	75

ENERGY

THE SUN-BAKED POTATO	76
NUCLEAR SOURCES	76
THE TEST TUBE GREENHOUSE	77
SOLAR SOURCES	77
START A FIRE WITH A MAGNIFYING GLASS	78
SOLAR SOURCES	78
WILL THE HEAVY BRICK HIT YOUR NOSE?	79
POTENTIAL VS KINETIC; THE PENDULUM	79
THE TWIN PENDULUM	80
ENERGY TRANSFER; THE PENDULUM	80

HEAT

THE BROKEN FLAME	81
CONDUCTION; KINDLING POINT	81
WHICH COIN WILL STAY ON LONGER?	82
RADIATION; ABSORPTION	82

MAGNETISM

THE MYSTERIOUSLY MOVING NEEDLE	83
MAGNETIC LINES OF FORCE	83
THE DROPPING RACE	84
ELECTROMAGNETIC INDUCTION	84

STATIC ELECTRICITY

THE ELECTRIC METER STICK	85
ATTRACTION OF UNCHARGED OBJECTS	85
THE LEVITATION ACT	86
STATIC CHARGES	86

CURRENT ELECTRICITY
    THE BALLOON FUSE 87
        CIRCUITS 87

LIGHT
    THE SIMPLE PERISCOPE 88
        REFLECTION 88
    SPEAR-FISHING, ANYONE? 89
        REFRACTION 89
    THE MAGIC OIL 90
        REFRACTION 90
    WHY DO WE SEE TWO COINS? 91
        REFRACTION; TOTAL REFLECTION 91
    USE WATER AS A MIRROR? 92
        REFRACTION; TOTAL REFLECTION 92
    THE REFLECTING BRICK WALL 93
        REFRACTION; TOTAL REFLECTION 93
    MAKE YOUR OWN RAINBOW 94
        REFRACTION & SCATTERING; TOTAL REFLECTION 94

SOUND
    PLUCK A RUBBER BAND 95
        PITCH 95
    HEAR THE DOPPLER EFFECT 96
        DOPPLER EFFECT 96
    THE ELLIPTLCAL WONDER 97
        REFLECTION 97

FORCES
    THE BURNING CANDLE SEESAW 98
        TORQUES; CENTER OF GRAVITY 98
    WHERE IS THE BALANCING POINT? 99
        CENTER OF GRAVITY 99
    THE PLATE CAROUSEL 100
        CENTER OF GRAVITY 100
    THE UNREACHABLE CUP 101
        CENTER OF GRAVITY; HUMAN BODY 101
    STUCK TO THE WALL? 102
        CENTER OF GRAVITY; HUMAN BODY 102
    THE CENTER-SEEKING PAPER 103
        CENTER OF GRAVITY; ADHESION 103
    THE STANDING MATCHBOX 104
        CENTER OF GRAVITY; SHOCK ABSORBERS 104
    STAND A DOLLAR BILL ON YOUR FINGER 105
        CENTER OF GRAVITY; AIR RESISTANCE 105
    KICK A STRAIGHT LINE 106
        CENTER OF GRAVITY; SPINNING OBJECTS 106

STAND A RAW EGG ON ITS HEAD 107
    CENTER OF GRAVITY; SPINNING OBJECTS 107
THE MAGIC STRIP OF NEWSPAPER 108
    ADHESIVE FORCES; FRICTION 108
THE INVISIBLE GLUE. 109
    FRICTION 109
THE CARDBOARD BOTTOM. 110
    WATER PRESSURE 110
THE SQUIRTING WATER HOLES.. 111
    WATER PRESSURE 111
THE OUTPOUR RACE... 112
    WATER PRESSURE; VORTEX IN A LIQUID 112
HIT THE BOTTLE ON THE BACK SWING?.. 113
    THE PENDULUM 113
HOW MANY SWINGS CAN YOU GET?.. 114
    THE PENDULUM 114
IS THE BALL REPELLED?... 115
    ELASTICITY; CONSERVATION OF MOMENTUM 115
PIERCE A POTATO WITH A STRAW?... 116
    RIGIDITY 116
THE LOOSE KNIFE SUPPORTS... 117
    STRENGTH OF WEAVED MATERIAL 117
THE DOLLAR BILL BRIDGE... 118
    STRENGTH OF CORRUGATED MATERIAL 118
HOW LONG CAN YOU HOLD THE BURNING PAPER?... 119
    STRESSES IN PAPER 119
THE STICKY KNIFE... 120
    STACKING FORCES 120
HOW CAN WE GET THE CORK OUT? (I)... 121
    NEWTON'S THIRD LAW 121
HOW CAN WE GET THE CORK OUT? (II).. 122
    FRICTION.. 122
THE WATER HAMMER... 123
    COMPRESSION; WATER PRESSURE 123
PUSH A NEEDLE THROUGH A BALLOON?.. 124
    SHEARING 124
THE WATER-TIGHT ZIPLOCK BAG... 125
    STRETCHABILITY 125
THE MYSTERIOUSLY MOVING BALL 126
    FREE FALL; CENTER OF GRAVITY 126
CRUSH THE CAN BY STANDING ON IT?.. 127
    PERPENDICULAR FORCES 127
THE CONFUSED TWIRLING PAPER 128
    ROTATIONAL FORCES; TORQUES 128
THE YIP-YIP STICK 129
    ROTATIONAL VIBRATIONS; DIRECTIONAL OSCILLATIONS 129

CRACK A NUT	130
PRESSURE VS FORCE	130

SPACE SCIENCE

ANOTHER BALANCING ACT	131
NEWTON'S FIRST LAW	131
PULL THE DOLLAR BILL OUT?!	132
NEWTON'S FIRST LAW	132
GET THE EGG IN THE GLASS	133
INERTIA; NEWTON'S FIRST LAW	133
THE STRAW ROCKET	134
NEWTON'S THIRD LAW	134
THE GRAVITY MACHINE (I)	135
CENTRIPETAL FORCE; GRAVITATION	135
THE GRAVITY MACHINE (II)	136
CENTRIPETAL FORCE; GRAVITATION	136
HOW HIGH WILL THE BALL BOUNCE?	137
CONSERVATION OF MOMENTUM	137
THE TEST TUBE CANNON	138
CONSERVATION OF MOMENTUM; RECOIL	138
THE SPINNING FOOTBALL	139
ANGULAR MOMENTUM; MOMENT OF INERTIA	139
THE HUMAN GYROSCOPE	140
ANGULAR MOMENTUM; PRESESSION	140
THE TIN CAN RACE	141
ANGULAR MOMENTUM; MOMENT OF INERTIA	141
THE CUP OF COFFEE DROP	142
WEIGHTLESSNESS	142
THE WEIGHTLESS NAIL	143
WEIGHTLESSNESS	143

EARTH SCIENCE

WHAT IS THE ROCK'S S.G.?	144
SPECIFIC GRAVITY	144
HOW CAN WE DETERMINE THE ROCK'S VOLUME?	145
ROCKS & DENSITY	145
SIMULATE A VOLCANO ERUPTION	146
VOLANIC ERUPTION	146

PLANTS

THE SWOLLEN EGG	147
GROWTH; OSMOSIS	147
THE TOOTHPICK STAR	148
CAPILLARY ACTION; ABSORPTION	148

HUMAN BIOLOGY

ARE WE PARTIALLY BLIND?	149

ARE WE PARTIALLY BLIND? 149
    THE HUMAN EYE 149
ARE YOU LEFT- OR RIGHT-SIGHTED? (II) 150
    EYESIGHT 150
THE SWAYING CARDBOARD 151
    PERCEPTION 151
THE FLOATLNG PIECE OF FINGER 152
    PERCEPTION; EYESIGHT 152
THE HAND IS QUICKER THAN THE EYE 153
    ILLUSION; EYESIGHT 153
PUT THE BIRD IN THE CAGE 154
    ILLUSION; EYESIGHT 154
THE ELLIPTICAL PENDULUM SWING 155
    ILLUSION; EYESIGHT 155
IS THE WATER WARM OR CQLD? 156
    NERVOUS SYSTEM; TOUCH 156
CATCH THE DOLLAR BILL 157
    NERVOUS SYSTEM; REACTION TIME 157
HOW FAST CAN YOU REACT? 158
    NERVOUS SYSTEM; REACTION TIME 158
THE UNCONTROLLABLE FOOT 159
    MUSCLE COORDINATION 159
THE KICKING FROG LEG 160
    MUSCLE COORDINATION 160
ARE WOMEN MORE AGILE THEN MEN? 161
    BODY BUILD OF MEN VS WOMEN 161

INDEX 162

- # Be a Happy and

- # Enthusiastic Teacher!

INTRODUCTION

A point of concern that dedicated science teachers will sooner or later face in their career, is how to arouse the students' interest in the subject matter that he or she is teaching. This is the age old problem that will confront every single teacher, whether at the primary, the secondary, the university, or at the post graduate level. It is a problem of motivation. The question is: how do we teachers arouse the students' interest? In modern terminology: how do we turn the students on to whatever subject matter we teach, in our case science?

How do we turn someone on to learn to play the piano? - or a toddler to eat his/her meal? - or anybody to learn to swim? Harry Overstreet, in his illuminating book: **Influencing Human Behavior,** said: *"Action springs out of what we fundamentally desire.... and the best piece of advice which can be given to would-be persuaders, which you and I are, whether it is in the home, in business, in school or in politics, is: First arouse in the other person an eager want. He who can do this will have the whole world with him. He who cannot, will walk the lonely way."*

We certainly do not want to be placed in the latter category. How can we then create this **eager want,** in our case an eager want in the student to learn? The answers to the above questions are thus quite obvious. By showing and letting the person listen to beautiful piano playing we arouse in the other person a want to play the piano. By showing the toddler that you yourself enjoy eating whatever food you want him/her to eat, an **eager want** to eat is created in the young person. By showing that swimming is a lot of fun, **the want to learn** to swim is aroused in the other person.

Similarly, an **eager want** to learn science can be created in the student by showing the student that doing science can be very enjoyable. Unless there is a need, or a want in the person, or a reward for the other person, we cannot turn anybody on to do anything. Romey 1) stated that "the science teacher's first task is to **establish a need to know"**. How can you make someone try to taste a foreign dish? Would you just describe the dish or force the person to take a bite? Neither! Unless you yourself show the other person that you enjoy eating the particular dish and really show it (smacking your lips accompanied with exclamations of ecstasy!), that person most likely won't touch the strange dish.

So it goes with learning science or any other subject matter. In order to get students **fired up** about something, first you yourself have to be **enthused** about it. Like in a wood fire, in order for you to give others a spark, first you yourself have to be aglow and burning. Thus, showing **enthusiasm,** to be dealt with under **Science Teacher's Characteristics**, is one of the conditions for good teaching.

In the process of teaching and learning of science in the classroom, and the interaction between teacher and student, in order for the teaching to be effective and for the learning to be optimal, several key questions need to be answered:

1) What are the characteristics of the really successful and outstanding science teachers?
2) What is the nature of science itself as subject matter to be learned?
3) What are the characteristics of the learner and the conditions that have to be met in order for learning to take place?
4) Is there a specific device, method or technique that would create the **need to know** or the **eager want** to learn in the student?

This introductory section attempts to bring out the answers to the above questions in a logical progression, with the assumption that you, the reader, are already motivated yourself to read further.

1) Science Teacher's Characteristics

The features that will be discussed here are the special attitudes of a teacher that were seldom encountered in the average teacher, but that were always present in successful and effective science teachers.

a) Enthusiasm

Do you show any enthusiasm in your own science teaching? Have you ever asked yourself whether the degree of your enthusiasm in science that you show in class is high enough for it to spill over to your students? High enough to turn your students on to science?

Several works have been published on the subject of **enthusiasm.** One is by the author of "The Power of Positive Thinking", Norman Vincent Peale's [2]: "Enthusiasm Makes the Difference", in which he states: *"It is a proven law of human nature that as you imagine yourself to be and as you act on the assumption that you are what you see yourself as being, you will in time strongly tend to become, provided you persevere in the process."* In other words, if you want to be enthusiastic, think yourself already possessing all those characteristics of someone that is enthusiastic and **act enthusiastic!** And you'll eventually be enthusiastic!

If you think back of your own education and ask yourself how you became interested in science or how you became a science teacher, you almost certainly would be able to trace it back to one or two teachers, whether it was at the elementary, secondary or college level, who were enthusiastic enough in science so that he/she made you want to continue a career in that field.

Enthusiasm is a funny thing. Although you cannot teach it to the students, they can actually catch it! *"Enthusiasm, like measles, mumps and the common cold, is highly contagious"*, says Emory Ward. One of the most enthusiastic teachers of science I know is Hubert Alyea [3], originator of T.O.P.S. (Tested Overhead Projection Series) and Armchair Chemistry. His boundless zest, self-reliance and overflowing enthusiasm for chemistry teaching is very contagious. As a graduate student I once visited him at Princeton University to interview him about TOPS. Although he was very busy at that time, he made a special effort and insisted that I should see one demonstration that he would like to do for me. This was the clock reaction where a colorless liquid would turn to orange and blue on command! He did it with such enthusiasm that I caught some of it and could actually weave some zest into my own presentations to pre- and in-service science teachers.

Another enthusiastic science educator is Vernon Rockcastle, Professor of Science Education at Cornell University. His vitality and zest for science teaching has made him not only an author of science texts, but also one of the most outstanding science education presenters at conferences. His science workshops offered at professional meetings are always some of the most enjoyable ones to attend.

Other science educators who stand out in my mind are: Paul F. Brandwein, researcher and author of numerous Science Methods textbooks; Harry K. Wong, Junior High School Science Teacher, author, and most sought after speaker in the science education world; Irwin Talesnick, Chemistry Professor at Queens University and most dynamic demonstrator of chemistry; George VanderKuur, scientist and physics educator at the Ontario Science Centre; Alan McCormack, Professor of Science Education and Zoology at the University of Wyoming; and many, many more which are too numerous to mention.

What is a common characteristic that all these zestful and dynamic speakers possess? It is **enthusiasm!** Moreover, they are not afraid or inhibited to show it. They actually move faster, talk faster, smile more, and have a good time themselves! What are other features that they have in common?

b) A High Self-esteem

Every single science educator or teacher mentioned above possesses a **high self-esteem.** They feel good about themselves. They value themselves highly. They know that they have something of value to contribute and this shows in their stance, poise, manner, and the voice volume and intonation that they use. In other words, their **attitude** is right!

How can you **develop the right attitude** as a science teacher? Attitudes not only show through but also "sound" through to the students. When you come in the classroom, it is not only your clothing that students will look at, but they will also listen to the voice and tone that you are using, which all determine your attitude. Expressions, voice volume, and inflections that you are using will give away your attitude towards science and science teaching in general. It will tell the students whether you like science, whether you are truly enjoying your occupation as a science teacher.

David J. Schwartz [4], in his book The Magic of Thinking Big suggests to make your attitudes your allies in everything you do. In the chapter that has the same title as the suggestion, he talks about growing the following three attitudes to win for you in every situation:

i. Develop the Attitude of "I'm Activated"

Let me illustrate this with the chemistry class that I was enrolled in at the Grade 12 level. At that point I actually already developed an interest in chemistry because of the previous years' chemistry courses, which were offered by rather good teachers. Then came this grade 12 teacher (Mr. A), who was a middle-aged gentleman, apparently knowing the chemistry content very well, but nevertheless, was pathetically dull. He wrote constantly on the board all the reactions that he talked about, but never showed us any demonstration. It was amazing how he could change an exciting subject like chemistry into one that seemed so terribly dull and boring.

You may well imagine what the effect of such a boredom was on the students. Almost no one paid attention, some were sleeping and some were reading comic books, and actually almost none of the students learned anything.

Well, one day this teacher did not show up. As a matter of fact he completely disappeared from school - I really did not know why - but I think it was a lucky break for us students, because he could have turned everybody off in chemistry, actually he was already in the process of doing so. If he had stayed on, I know that I would certainly not be in the field of science education.

With the chemistry teacher missing, the grade 12 class had to go to the medical college, where a chemical engineer (Mr. B) was offering our grade 12 students special chemistry classes. This teacher then was completely the opposite of Mr. A. He was enthusiastic and he accompanied and illustrated every single reaction that he put on the board with a demonstration. How exciting! Chemistry became alive to me! Everybody was spellbound and our eyes were glued to the teacher and what he did. Questions were invited at any time and discussions were quite frequent. Everybody was learning!

The most remarkable thing is that I still remember some of the details of that course and Mr. B himself. I can still remember (and this is 34 years ago) his face, his smile, and his name: Ir. (Engineer) Van Gyn; and the most outstanding demonstrations of spontaneous combustion reactions, most of which are described in this book under the chapter "Chemistry".

Can you see the difference between Mr. A and Mr. B? In the case of Mr. A the students had no interest whatsoever in what the teacher was saying, because the teacher himself was not interested - at least he did not show it - in the subject matter. He must have been bored with chemistry and it certainly showed through.

The second teacher, Mr. B, just was himself and showed interest in his subject matter. He showed to his students that he liked what he was doing, that he enjoyed himself, and voila - his students became interested! His enthusiasm spilled over and was caught by his students! Thus the lesson from these examples is: **TO ACTIVATE OTHERS, YOU MUST FIRST ACTIVATE YOURSELF!** In order to get people enthusiastic, you must first be enthusiastic yourself. Before you can give off a spark, you must first be aglow yourself.

ii. Develop the Attitude of "You are Important"

Every human being, no matter whether rich or poor, young or old, backwards or sharp, living in New York City or Timbuktu, has a desire to feel important. Man's most compelling, non-biological craving is this desire to be important.

John Dewey, one of America's most profound philosophers, said that the deepest urge in human nature is **"the desire to be important"**.

Consider the following two classes and imagine yourself as a student coming into these classes the first day of school:

Class A
The teacher starts the class by saying: "Boys and girls. I am your science teacher. I am the boss and you have nothing to say. You are here only to learn from me. Science is important for our future, that is why we are here to learn science."

Class B

The teacher starts the class by saying: "Ladies and gentlemen. I am your science teacher. You are our future generation. You will be the leaders of our country 10-20 years from now. Since science plays such an important role in our future, you are here to learn what science is all about."

In which of the two classes would you feel more at home? In which of the two classes would you feel more important? In which would you likely learn more? It is quite obvious, of course in class B. Why? Because the teacher made you feel that you will be the leader of the country and leaders must be knowledgeable about science.

Why do we want to make others feel important? Because not only will people do more for you, but when you help others feel important, you will help yourself feel important too.

In the classroom we teachers can do this by praising the students whenever they are doing the right thing, or whenever they respond with the correct answer. Give praise at every opportunity you have. Respond to correct answers with: "Excellent!" or "Bravo!". Build students up and make them feel that they are contributing valuable information to the discussion. By praising the students you make them feel good, and good feeling students will never give you trouble in class. It is those students that have either bad feelings towards you or those that hate to be in your class, that will give you discipline problems.

So MAKE YOUR STUDENTS FEEL IMPORTANT by praising them and looking at the positive things that they did. In doing so, the students will feel good and love to be in your class.

iii. Develop the Attitude of "I Care for You"

The third attitude to develop to get the right teacher's attitude is showing others, the students in this case, that you sincerely care for them. This attitude should be quite apparent to the students. They should be able to sense and feel that you actually love your students. Students should know that you, the teacher, are there to help the students to learn in any way you possibly can, because you love them.

This aspect of "love" is the most neglected one in the attitude of the teacher in our educational system, whether it is at the elementary, secondary, or at the college level. The kind of love we are talking about here is comparable to parental love, or brotherly/sisterly love (when the age difference between teacher and student is still small). The love of a parent that is totally and unequivocally dependable. The love that a child feels coming from a parent, that gets him/her through thick and thin in life.

Harry K. Wong [5] expresses his love to the students by actually touching them. The kind of touch he is talking about and uses are the following: a) shake hands, b) a pat on the back, and c) a punch against the shoulder. This kind of touch he uses to enforce his appreciation or approval after the student has achieved something or has done something correctly, accompanied with a comment like: "Hey Tom, I knew I could count on you!" or "That-a-girl Susan, that's the way to do it!", builds the students self-esteem.

Love in the classroom as expressed by the teacher for his/her students, is the unconditional acceptance of the students, whether they are yellow, white, red, or black, whether they have a 75 or 140 IQ. It is always "looking for the good" in the student and noticing the positive side of things.

Denis Waitley [6] in his book <u>Seeds of Greatness</u> states: "With love, there can be no fear. Love is natural and unconditional. Love asks no questions - neither preaching nor demanding; neither comparing nor measuring. Love is - pure and simple - the greatest value of all."

So bring love into your classroom, but before you can do that you must be able to totally accept yourself. Waitley calls this one of his best kept secrets of total success: "WE MUST FEEL LOVE INSIDE OURSELVES BEFORE WE CAN GIVE IT TO OTHERS"

In developing a healthy self-esteem we must understand that the word **esteem** means *to appreciate the value of*. We must realize that each of us are unique individuals and that **we are a masterpiece of creation.** Waitley says: *"There will never be a person who is more important than any other person, no matter how they look and no matter what kind of work they do. Each of us is as valuable and worthwhile as any other person."*

c) Creativity

Another common characteristic that the successful science teacher possesses is **creativity.** Among all living organisms on earth, only the human being is endowed with creativity or creative imagination. Napoleon once stated: "Imagination rules the world." Einstein said: "Imagination is more important than knowledge, for knowledge is limited to all we now know and understand, while imagination embraces the entire world, and all there ever will be to know and understand."

A study that was recently conducted by a research team of Stanford University revealed that: *"what we watch is having an effect on our imaginations, our learning patterns, and our behaviors"*. This means that we have to be selective in what we feed our minds with. Our mind is like a computer: feed it with good stuff and good stuff will come out of our mouths. In the computer world there is a term: GIGO meaning - garbage in, garbage out. Feed a computer with garbage and you can't get more than garbage out of it.

Denis Waitley [7] has written a whole chapter on creativity related to success in life. He mentioned a few characteristics of creative individuals. How many of them fit your personality?

* Optimistic about the future
* Constructive discontent with status quo
* Highly curious and observant
* Open to alternatives
* Daydreamer, projecting into future
* Adventurous, with multiple interests
* Ability to recognize and break bad habits
* Independent thinker
* Whole-brain thinker (innovative ideas into practical solutions)

Have you met any individuals with most of the above mentioned features? If so, make friends with them, make them your role models, associate with them, and learn as much as you can from them.

d) Responsibility

Most teachers do possess this characteristic, which is one of the most valuable features for a teacher to have in the teaching world and actually for anyone to have in life in general. A responsible person is dependable, can be trusted, is a person in whom other people have faith. In other words, his word is to be trusted. When he says that he'll be present at a certain place for an appointment at such and such a time, he should be there at that particular time, unless he has some legitimate reason for not showing up. This goes especially for business people in the business world. Can you imagine a business man making a very important appointment and the next day he comes with the excuse: "Oh, I forgot!" Will he have credibility with his colleagues? Will his colleagues put trust in his words?

What does this mean for the teaching world? In teaching the teacher's word should be believed by the students. The teacher should be consistent and should actually do what he says he would do, whether it's in giving back papers or whether it is in carrying out penalties for not doing homework, or any other thing. It is just a matter of remembering what you yourself said (or promised) and carrying out/performing what you said. This means that **one should actually do what one preaches.**

Many teachers (especially at the college level) say one thing and do another. For instance, one might be lecturing about the excellent advantages/effectiveness of the *activity approach* in teaching science, but one keeps teaching the students science by lecturing! This way, the whole thrust of the course might be in jeopardy. Students immediately lose trust in the teacher. Students will not put any value to the ideas that the teacher may have to offer. By not putting into practice what you, as a teacher, are saying - in theory -, your preaching about the theory is mostly in vain!

Another aspect of this "doing what you preach" is the role model part. You, the teacher, are usually the students' role model. You give the example to the students in your whole attitude towards science and learning, and in living your life in general. So, if you want your students to develop responsibility, you yourself have to demonstrate all the desired characteristics of a responsible, dependable personel of high integrity. Thus, your motto should be: **"Never preach what you don't practice!"**

In order to develop responsibility in your students, you need to give them opportunities to do so. Let them show you that they can be responsible individuals. Give them tasks that demand some responsibility, like organizing books in class, cleaning and organizing glassware and materials in the laboratory, making posters, taking care of animals and plants, etc. The more we do for the students things that we like for them to be able to do, the less they can do those for themselves. As a father of four children and a teacher of teachers, I know from experience that the most lasting gifts that parents can give to their children, or that teachers can give to their students, are **roots and wings: Roots of Responsibility and Wings of Independence.** When you, as a teacher, can leave your students on their own, and learning is still continuing, then you have actually taught your students responsibility and independence.

e) A Sense of Humor

A teacher should have a good sense of humor. This does not mean that he must be a comedian or constantly have to come up with funny stories. A good sense of humor means that the teacher must be able to look at his own mistakes and laugh at himself.

Before you can laugh at yourself and the mistakes you made, you have to be able to accept and admit that you have made a mistake. We are all human beings and being human means that we all are bound to make mistakes. By admitting your mistakes in front of the students you actually show them that you, as a teacher, are human too. Students would rather see you admit your mistakes than see you try to cover them up.

Not only should you be able to admit your mistakes, but also the answers to questions that you don't know! Paul Brandwein once stated that teachers should not say: "I don't know", they should say: "It's not in my field". He was of course just joking when he made that statement. You may comfortably say: "I don't know" and the students do expect you to say it to some of the questions that they pose, but you might be able to suggest where they can look for the answer. You see, a teacher is just another human being that knows a little bit more than the students, but he certainly is not a walking encyclopedia.

So be optimistic, put a smile on your face and get excited about the fact that you are so privileged to be able to determine the thinking in our younger generation and thus the future of our country!

f) Communication

Successful science teachers and educators are always also successful communicators. They seem to be able to capture the attention of their audiences, keep them awake for an extended period of time, and get across whatever concepts or messages they wanted their audiences to know.

In order to get things across, we may want to communicate in different languages if we are dealing with people of different nations. I was once in Europe and saw a sign in front of a restaurant. It said: "All Languages Spoken Here", so I stepped in and asked the waitress: "Who speaks all languages?" She said: "The customers!" But even in the same English language, it can sometimes be difficult to communicate well. Like this woman that wanted a divorce and saw her pastor. He asked her: "Do you have any grounds?" "Yes", she said, "We own about 15 acres North of town". "No, that is not what I meant", said the pastor. "I mean, do you have any grudge?" "No", she answered, "But we do have a carport in our driveway". "No, no", said the pastor, "How can I put it? Does your husband beat you up?" To this question she reacted: "Beat me up? I am up much earlier than he does every morning of the week!". Quite exasperated he said to the woman: "Lady, would you please listen carefully! Does your husband give you any problems?" "Yes", she said, "He just cannot communicate!"

Nido R. Qubein [8] in his book *"Communicate Like a Pro"* stated: ....**sincere, honest, enthusiastic, and positive people make the best communicators. The better you get to know them, the more you like and trust them; and the more you are willing to listen to what they have to say.** This certainly is true for the effective teacher and his success in communication with his students.

In science, however, above and beyond the general features above, successful teachers have a particular way of communicating a concept to the students. They do not only try to describe the concept with words, but they usually show the student an object or event which applies the particular concept. A **Law of Teaching Science** which Paul Brandwein [9] coined is:

**A CONCEPT IN SCIENCE IS SYNONYMOUS WITH
A CORRESPONDING SET OF OPERATIONS.**

This means that whenever we want to convey or communicate or teach a science concept to the students, we should show or demonstrate a corresponding set of operations in which that concept is applied. In other words, we should do a science demonstration or involve the students in an activity which applies the concept or principle.

In science, therefore, (and in my opinion even in other disciplines) if you want to teach a concept, do also a demonstration which applies the concept. This is why the demonstration is such an important tool for the science teacher. A shorter version of Brandwein's Law is:

DEMONSTRATION IS TEACHING

SUMMARY: SCIENCE TEACHER'S CHARACTERISTICS

a) **DEVELOP ENTHUSIASM**

b) **DEVELOP A HIGH SELF-ESTEEM**
  i. DEVELOP THE ATTITUDE OF "I'M ACTIVATED"
  ii. DEVELOP THE ATTITUDE OF "YOU ARE IMPORTANT"
  iii. DEVELOP THE ATTITUDE OF "I CARE FOR YOU"

c) **BE CREATIVE**

d) **SHOW RESPONSIBILITY**

e) **DEVELOP A GOOD SENSE OF HUMOR**

f) **DEVELOP YOUR COMMUNICATION SKILLS**

2) The Nature of Science and the Scientist

In order to teach science properly, we have to know what particular characteristics science has, and also the particular features that scientists display.

a) The Nature of Science

If we look in Webster's Dictionary, there are several definitions for science, but the closest one to what teachers are concerned with in school is this: "Science is a branch of study that is concerned with observation and classification of facts and especially with the establishment or strictly with the quantitative formulation of verifiable general laws chiefly by induction and hypotheses".

The science philosopher Benjamin [10] defined science as: **"That *mode of inquiry* which attempts to arrive at information about our world (universe) by the method of observation and by the method of confirmed hypotheses based on observation."**

In both definitions we can see that science is a process as well as a product. In the latter one we can see more of an activity rather than just the study of facts. If we hear the word "inquiry", what do we immediately think of? A process of questioning and answering to search for the truth of the matter. It is that part of the inquiry process which tries to obtain truthful information about our universe by using our senses, by gathering data, by formulating hypotheses, and by confirming these hypotheses on the basis of what we obtained from observation.

What are the implications in teaching this particular discipline in the classroom? The best way to teach a particular subject is to reflect the true nature of that subject. In our case, reflecting the nature of science is by doing science activities in the classroom. By having the students involved in observing, measuring, calculating, formulating hypotheses, gathering data, etc. in other words, involving them in all the science processes.

In dealing with the processes, it is virtually impossible to avoid dealing with science concepts. We have to observe some event, hypothesize why something is happening. Put in other words: there has to be a focus or science concept around which students can practice the processes. This book provides the teacher with the needed focuses - **discrepant events** - which are interest arousing and challenging.

By presenting these **discrepant events** to the students in the proper way (see section under the discrepant event), it is very natural and logical for the teacher to involve the student in the process of inquiry. Since the event is so unexpected or against our intuition, it is easy for the teacher to ask questions about it. The teacher can ask questions requiring either simple answers or more complicated answers, depending on how deep he wants to go into the subject. The teacher can therefore adjust each of the **discrepant events** to his own teaching level.

b) The Nature of the Scientist

We teachers are especially interested in what makes a scientist work on one topic with such dedication. How do scientists become curious about a particular topic and start a series of investigations about it that could sometimes last for a whole lifetime? What is it that triggers their attention to such a degree that they can become so engrossed in it? In other words: what is it that makes them tick?

Let us consider a couple of great discoveries and let us take a closer look at their discoverers. Just before the turn of the century Henry Becquerel discovered radioactivity and a few decades after the turn of the century Sir Alexander Fleming discovered penicillin.

Becquerel was fascinated by Roentgen's discovery of X-rays and wondered if the natural luminescence, or glow, of certain minerals might also be accompanied by similar X-ray emission. He took a thin crust of a uranium crystal and placed it on a photographic plate wrapped in lightproof paper. The wrapping was exposed to sunlight for several hours. When developed, an outline of the crystals showed up on the photographic plate. Subsequently, in a similar experiment he found that where some uranium crystals had been left in a dark drawer, unexposed to light, together with a photographic plate, they left even a stronger outline on the developed plate! At that moment he must have been very surprised and could have exclaimed: "Hey! There is a **discrepancy** here!" He became so interested in this phenomenon that during the following years he became totally engrossed in his work on this strange radiation - **radioactivity** !

Fleming was working in the laboratory with cultures of Staphylococci, bacteria that causes boils and blood poisoning. He had also cultures of Penicillium mould around and one day some of his bacteria cultures were contaminated with the Penicillium. While he was in the process of discarding the contaminated bacteria cultures his attention and interest were aroused by **something unusual** : there were certain spots in the bacteria cultures that were blank. In other words, the poisoning causing bacteria did not grow where it was contaminated with Penicillium. From that moment on the investigation and further extraction of penicillin from the penicillium mould continued.

From the above examples we can see that in both cases the major cause that triggered the scientists' attention and interest was a **discrepancy.** Something that they did not expect to happen at all, something that is against their intuition. For most people in general, something that is out of the ordinary may on the whole attract one's attention. Also something that is unusual generally can be remembered for a longer period of time.

What are the implications of those phenomena or happenings in the teaching of science and in making the subject matter more interesting to the students? If we consider the scientists and how they became so interested in the particular topics, it was a **discrepant event** that triggered it. Similarly, we can confront the students with events that are **discrepant to them,** so that their curiosity and interest is aroused. These discrepant events can be carried out with relatively simple materials. The main characteristic that you, the teacher, should show in demonstrating a discrepant event to the students, is to show your own excitement or surprise in front of them. Although that particular discrepant event might not be discrepant to you anymore, you still should show your own perplexed expression. The students should see your own astonishment and surprise in your facial expression!

Confronting the students with **discrepant events** and meeting all the conditions for the learning process would be the ideal things that teachers should remember. This brings us to the question of which factors are actually having a pertinent influence on the learning process?
The two factors are:
  a) How can new information be retained longer by students?
  b) What are the conditions for information retention?

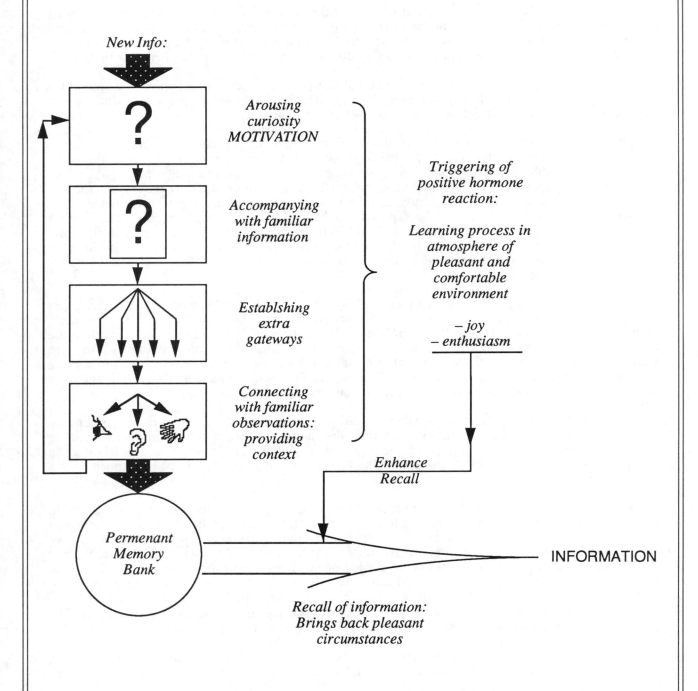

Invitations to Science Inquiry – Supplement – Page 12

3) The Learning Process

In considering the Learning Process or the Process of Information Retention as described by Vester [11], it can be seen that several conditions have to be met in order for new information to penetrate our Permanent Memory Bank (see flowsheet on page 12).

Vester mentioned the following conditions to be most essential in order for new information to penetrate the Permanent Memory Bank:

a) Arousing curiosity: MOTIVATION
b) Accompanying with familiar information.
c) Establishing extra gateways.
d) Connecting with familiar observations: providing context.
e) Providing an atmosphere of JOY and ENTHUSIASM.

Let us take a closer look at each of these conditions:

a) Arouse Curiosity

Of the above mentioned conditions, the most important one is the arousing of curiosity in the learner. Baez [12] mentioned that curiosity is one of the four important traits in people that should enrich the quality of life. He called them the four C's, which stands for: "Curiosity, Creativity, Competence, and Compassion". About curiosity he further states: "Curiosity is the motor that drives the scientist's curiosity; it is the source of discoveries in science and technology. The spark of curiosity ought to be fanned into a flame by teachers and parents. It can make learning a pleasurable experience, but it is sometimes stifled by uninspired teachers who find it easier to demand rote learning."

How can we fan the spark of curiosity into a flame? What can the science teacher do to arouse the students' curiosity? Romey stated that: "One of the best ways to stimulate interest is to **offend the student's intuition** in some way or to confront him with a situation that is not readily acceptable". Sund and Trowbridge [13] are of the opinion that..."another technique for developing motivation and interest in a discussion is to use **pictorial riddles**".

What better way to offend the student's intuition in science than by doing a demonstration? A demonstration of which the outcome is not expected or of which the performance is thought impossible: a demonstration of a **discrepant event**. Gagne [14] is of the opinion that in order for learning to occur, the student's **brain must be awake.** The learner must have a state of alertness that corresponds to the common sense word ATTENTION.

b) Use Simple Materials

The use of simple materials or accompanying the question with familiar information is also one of the essential conditions for the learning process. In order for students to learn something new, the teacher has to start with something that the student already knows and is familiar with.

Especially in presenting a demonstration to the student with the purpose of showing a **discrepant event,** it will not **be successful unless** the demonstration is carried out with simple materials which the student is familiar with. Brandwein [15] says: "Unless an object or event is recognized, a **problem is** not recognized". In other words, a discrepancy or discrepant event is not recognized unless an object or event is recognized."

c) **Use All Gateways**

With these we mean gateways into the human brain. There are of course, five gateways or five entrances into our brain, each opened by the use of our five senses: sight, hearing, touch, smell, and taste.

When teachers strictly lecture, they make use of only one gateway, which is the students hearing. As soon as teachers show or demonstrate something, sight is utilized as an entrance into the brain. But it is only when teachers give the students an opportunity to do things themselves, it is then that they also utilize the other three gateways into the brain: touch, smell, and taste. This means that when teachers illustrate their lectures with a **demonstration,** and then involve the students with a **follow-up activity** on the same concept, they make maximum use of gaining entrance into the students brain.

d) **Provide Context**

Any science concept or principle when taught isolated from our daily life, becomes completely meaningless and thus much harder to learn.

Therefore, **give examples** of where that particular concept presents itself in our daily life, or where that particular **principle applies.** Instead of talking only about the First Law of Newton, ask: "Why do you hold on to something solid when stepping into a bus that is just about to leave?" (First part of First Law) "Why do you do the same when the bus is about to stop?" (Second part of First Law).

It is when science concepts or principles are placed in context and connected with familiar experiences of the student, that they will become much more meaningful to the learner, and thus much easier to be learned.

e) **Do it with JOY and ENTHUSIASM**

Can you imagine yourself learning science from a teacher who thinks him/herself that science is boring? No, most likely not, as you probably would not learn any science, because it was portrayed as a boring discipline.

On the other hand, imagine yourself as a student coming into a classroom and seeing the teacher whistling to himself while preparing for a demonstration, jumping around to get the materials from the shelf here and from the shelf there, looking forward to conducting the lesson and just enjoying himself during the few minutes before the bell rings while waiting for the class to fill up. Would you expect to learn science in this classroom?

The answer to the above question is: "Of course!" The teacher s characteristic of having enthusiasm is therefore a very important one to possess. How do we acquire enthusiasm?

Not only is it a desirable teacher's characteristic, enthusiasm will also improve your personality. If you really want to change your personality, you can and it is not that hard to do! Here are a few excerpts from Norman Vincent Peale's *Enthusiasm Makes the Difference:*

>.................It is the change from apathy to enthusiasm, from indifference to exciting participation; it is an astonishing personality change which sensitizes the spirit, erases dullness and infuses the individual with a powerful motivation that activates enthusiasm and never allows it to run down.......
>.................You can develop enthusiasm, and of a type that is continuous and joyous in nature.......

................It has been established by repeated demonstration that a person can make of himself just about what he wants to, provided he wants to badly enough and correctly goes about doing it. A method for deliberately transforming yourself into whatever type of person you wish to be is first to decide specifically what particular characteristic you desire to possess and then to hold that image firmly in your consciousness. Second, proceed to develop it by acting as if you actually possessed the desired characteristic. And third, believe and repeatedly affirm that you are in the process of self-creating the quality you have undertaken to develop........
................It is a proven law of human nature that as you imagine yourself to be and as you act on the assumption that you are what you see yourself as being, you will in time strongly tend to become, provided you persevere in the process.

From the last two paragraphs we can conclude the following: **If you want to be enthusiastic, act enthusiastic!**

SUMMARY: CONDITIONS FOR LEARNING

Remember the acronym: **INSIGHT**

IN = INTEREST; DEVELOP INTEREST, AROUSE CURIOSITY

SI = SIMPLE; USE SIMPLE MATERIALS

G = GATEWAYS; USE AS MANY GATEWAYS AS POSSIBLE

H = HINGE; HINGE IT WITH EXAMPLES AND PUT IT IN CONTEXT

T = TIE; TIE IT ALL TOGETHER WITH JOY AND ENTHUSIASM!

4) The Discrepant Event

When properly used and presented to students, the **discrepant event** will encompass all five conditions for learning.

a) Psychological Background

The **discrepant event** in science finds its base in the Theory of Cognitive Dissonance by Leon Festinger [16], in which he proposes two basic hypotheses:

1. The existence of dissonance being psychologically uncomfortable, will motivate the person to try to reduce the dissonance and achieve consonance;

2. When dissonance is present, in addition to trying to reduce it, the person will actively avoid situations and information which would likely increase the dissonance.

It is especially the first statement that applies to using discrepant events to initiate learning. Festinger is further of the opinion that the **discrepant event** functions by causing dissonance between what is physically observed to occur and what one thinks should occur. Since it is impossible to change what is physically observed to occur already, the only alternative is to begin to seek information which will logically explain the occurrence.

Waetjen [17] stated that the **discrepant event** is a dissonant situation which results in arousal of conflict with a consequent need for the learner to assimilate or articulate the unknown, incongruous, unfamiliar material into his/her cognitive structure. To do this, he/she engages in exploratory behavior.

The **discrepant event** has also been called Disconfirmation of Expectancy. It is a situation contrary to what the learner expects, and actually a state necessary for cognitive development, for it moves the learner to try to resolve the discrepancy between what was expected and what actually happened.

It has also been called Conceptual Conflict. When the learner faces a situation which is in conflict with what he expects, the doubt, perplexity, contradiction, and incongruity play an important role in stimulating the learner's curiosity.

Piaget [18] says that: "This state (of perplexity and doubt) is a necessary first step in learning!", and further states that "puzzles are excellent sources for learning simply because they unsettle the learner, upset his intellectual equilibrium, and incite him to change or adapt his existing intellectual scheme. The learner who meets such a challenge develops and assimilates new skills that make him or her a cognitively richer individual."

The **discrepant event** then, whether it is a demonstration by the teacher or an activity performed by the student, is mainly used for its motivational effect on the learner. It creates the **eager want** in the student to know more about the event, and thus do other activities that are aimed at dissonance reduction; subsequent knowledge-seeking behavior is easily aroused and strongly reinforced.

b) What is a Discrepant Event?

Everybody has seen objects fall downwards. The fact is hardly surprising or unusual. However, if one were to see an object fall upwards, it would be an entirely different matter. It would be an event which defies gravity, and in this case the moving of the test tube against the force of gravity is a **discrepancy** (See The Upwards Falling Test Tube - ISI 2nd Ed. Event 1.9.).

Most people know that men are stronger than women. However, if one were to see men not being able to pick up a chair from a bent position and women doing it with ease, what would one think? Some feeling of surprise and curiosity would be aroused. This would especially be true when one gets involved in trying to perform the event oneself (See Are Women Stronger Than Men?).

All children know that a newspaper is so light in weight that it can be easily blown away by the wind. However, if one sheet of a newspaper placed on a long flat wooden stick (protruding over the edge of the table) would hold the stick down, even if one would hit the protruding end of the stick, what thoughts are being aroused in one's mind? (See The Heavy Newspaper).

One would expect that objects would fall down and not upwards, men are usually stronger than women and the chair is certainly not so heavy that men would not be able to lift it. One sheet of newspaper would surely not be heavy enough to hold down a stick, even to breaking! These examples of **discrepant events** are often described as **surprising, counter-intuitive, unexpected, paradoxical, mind capturing and intuition-offending.**

The **discrepant event** has the tendency to arouse strong feelings within the observer or the participant. Generally, one will have a sudden urge of wanting to know more about the event. One gets curious and wishes to solve the discrepancy in one's mind. Children as well as adults will demonstrate a strong desire to resolve the unexpected. The enthusiasm of the first will certainly be much greater. Children will simply not stop asking questions until they find out why certain events occurred the way they did!

When students interest is thus hightened, he/she will more likely become motivated to learn. **Discrepant Events** are therefore excellent means to create in the student an **eager want** to learn more about science. **Discrepant Events** capitalize on the students own curiosity, already present within the person, helping him/her to gain a better understanding and retention of science concepts.

c) **How to Use Discrepant Events?**

In order the achieve an atmosphere of inquiry in the classroom, it is important that the teacher presents the discrepant event as a science problem to be investigated or as a puzzle to be solved. The discrepant event or intuition-offending demonstration must be presented in such a way that the science principle or concept underlying the event is not immediately revealed. The teacher using the discrepant event should in general lines follow the procedure below:

Teaching Procedure in Using Discrepant Events

1. Presentation
   Present to the learner or involve the learner with the discrepant event by describing or commenting on the names of objects and operations only and not mentioning the reasons for the occurrence. In other words, the teacher may tell the student what he/she is doing and what materials he/she is handling, but not why something is happening.

2. Interaction
   Ask the learner questions that eventually will lead him/her to the main reason for the occurrence. In doing this the students will be engaged in science inquiry and actually practicing the science processes of observing, measuring, inferring, predicting, interpreting data, identifying and controlling variables, hypothesizing, and experimenting.

3. Involvement
   Participate the learner in similar (and simpler) discrepant events or counterintuitive activities illustrating and based on the same science concept. This will reinforce the learning and retention of that particular concept. Students may work individually or in pairs, or in groups depending on the availability of materials.

It is essential that the presentation of the intuition-offending demonstration or the involvement of the students in the discrepant event takes place under the following conditions (see also The Learning Process on page 13):

1. Arouse Interest
   The event should confront the observer with a perplexing problem. It should be presented almost like the way a magician would do a magic act.

2. Use Simple Materials
   The event should be performed with materials that are familiar to the learner, in other words use simple daily life materials.

3. Use All Gateways
   The students should have the opportunity to observe as well as carry out the events themselves. They should be allowed to totally experience the discrepant event.

4. Hinge It With Examples
   When dealing with the concept underlying the event, examples and applications of the concept in our daily life should be mentioned to make it more meaningful to the learner.

5. Tie It With Joy And Enthusiasm
   The teacher should show genuine enthusiasm in presenting the puzzling event and reveal his/her enjoyment in the subject matter in general.

When all of the above elements are present during the learning process, the student will be able to store or retain new information with much more ease. The learning experience becomes more meaningful and enjoyable, resulting in a much better retention of the science concept. The recalling process of the stored information will bring back all the pleasant circumstances during which it was learned, which makes it a nearly effortless mental activity.

The above conditions for learning a science concept may be attained when the teacher is genuinely showing concern for the students, uses discrepant events to initiate science inquiry, and most importantly is enthused about what he/she is doing. Once this optimum condition is reached, teacher and students will enjoy the teaching-learning process to a much greater extent.

d) The Empirical Evidence

Suchman [19] indicated in his study that the teaching technique with discrepant events positively influence content achievement. The study showed that this content achievement is affected by the arousal of student motivation, which is directed at the comprehension of the causes of the observed events.

Marlins [20] findings were confirmed by Liem [21], that "upper elementary school students taught using the demonstration-discussion method with counterintuitive events, had significantly higher retention scores compared to those taught without counterintuitive events."

I followed up my studies on the effect of discrepant events on science concept retention at the junior high level [22]. I subjected one group of students (control) to only discussion and reading of a text containing descriptions of discrepant events positively influences content achievement. The study showed that this content achievement is affected by the arousal of student motivation, which is directed at the comprehension of the causes of the observed events. Measures were administered to both groups:

The Pretest: administered before the lessons.
Post-test A: administered immediately after the lessons.
Post-test B: administered one month after the lessons.
Post-test C: administered three months after the lessons.

The results and means of the tests were then compared and subjected to the t-test described by Best [23]. The following t-values were obtained: Pretest: 0.27; Test A: 1.96; Test B: 1.78; Test C: 2.72. This shows that there was a significant difference between the means of Tests A and B at the .05 level, and of Test c at the .01 level, and also that there was no significant difference between the means of the Pretests.

From these findings one would be led to the conclusion that the larger the time lapse between the lessons and the administering of the test, the more significant the difference becomes between the mean score of the Experimental compared to that of the Control Group. In other words, the group taught using the discussion method whereby discrepant events were demonstrated by the teacher and experienced by the students, retained the science concepts longer compared to the group taught using the discussion method whereby discrepant events were only read.

The findings of the studies conducted thus far would seem to justify the use of **discrepant events** for teacher demonstration and student activities to increase retention and understanding of science concepts at the elementary and intermediate levels. One is also led to infer that the **discrepant events** lose their **discrepancy or motivating effect on the student,** when they are merely described in a text and read by the student.

THE FORMAT

This book is organized in such a way that it offers to teachers the optimum guidance for teaching science concepts and engaging the students in science inquiry. The chapters of the book are organized under an introductory section and four sections: Environment, Energy, Space and Forces, and Living Things. The introductory section provides the teacher with the psychological background for effective science teaching and the use of **discrepant events.** Each of the latter four sections is preceded by a short introduction providing the teacher with some information on the science concepts to be dealt with in the chapters.

The discrepant events, making up the main body of each chapter, are preceded by a list of objectives. These are stated in behavioral terms in order that student and teacher may work towards achieving these behaviors.

The discrepant events are invitations to science inquiry, which are all fitted on one page each, for simplicity in format and easy location. Each event is consistently constructed in the following format:

1. Upper left corner: Chapter Science Concept.
2. Upper right corner: Science Sub-concept or Science Properties.
3. Intriguing, curiosity arousing title of event.
4. Materials needed for the demonstration or activity.
5. Sketch of materials set-up.
6. Procedure, in step by step fashion.
7. Questions for teacher's use during demonstration.
8. Explanation, providing background information for the teacher.

The book ends with a list of Further Readings (First & Second Edition) in which similar or more detailed information may be obtained. A general index follows the reference list, where users of the book may find concepts and sub-concepts and the relevant events that are appropriate to use in teaching those concepts to students.

REFERENCES

1. Romey, W.D., INQUIRY TECHNIQUES FOR TEACHING SCIENCE, Prentice-Hall, Inc., 1968, p 16.

2. Peale, Norman Vincent, ENTHUSIASM MAKES THE DIFFERENCE, Fawcett Crest, New York, 1982.

3. Alyea, Hubert, TESTED OVERHEAD PROJECTION SERIES, & Dutton, Frederic B., TESTED DEMONSTRATIONS IN CHEMISTRY, J. of Chem. Ed., Easton, Penn. 1965.

4. Schwartz, David J., THE MAGIC OF THINKING BIG, Audio Tape, Sound Ideas, Simon & Schuster Audio Publishing Division, 1230 Ave of the Americas, New York, N.Y. 10020

5. Wong, Harry K., HOW YOU CAN BE A SUPER SUCCESSFUL TEACHER, Cassette Audio Tape Series, 1536 Oueenstown Court, Sunnyvale, CA 94087.

6. Waitley, Denis, SEEDS OF GREATNESS, Fleming H. Revell Co., 1983.

7. Ibid., pp 43-61.

8. Qubein, Nido R., COMMUNICATE LIKE A PRO, Berkley Books, NY 1983.

9. Brandwein, Paul F., THE METHOD OF INTELLIGENCE, Presentation to NSTA Convention, Toronto, 1968.

10. Benjamin, Abram Cornelius, SCIENCE TECHNOLOGY AND HUMAN VALUES, Columbia, Univ. of Missouri Press. 1965.

11. Vester, Frederic, HOE WIJ DENKEN, LEREN EN VERGETEN (Original title: DENKEN, LERNEN, VERGESSEN. Deutsche Verlags Anstalt GmbH, Stuttgart, 1975), Bosch en Keaning, Baarn, 1976.

12. Baez, Albert V., "Curiosity, Creativity, Competence and Compassion Guidelines for Science Education in the Year 2000", WORLD TRENDS IN SCIENCE EDUCATION, Atlantic Institute of Education, Halifax, NS, 1980.

13. Sund, R.B. and Trowbridge, L.W., TEACHING SCIENCE BY INQUIRY IN THE SECONDARY SCHOOL, 2nd ed., Charles E. Merrill, 1973.

14. Gagne, R.M., THE CONDITIONS OF LEARNING, Holt, Rinehart and Winston, Inc., New York, 1965.

15. Brandwein, Paul F., THE METHOD OF INTELLIGENCE, Presentation to NSTA Convention, Toronto, 1968.

16. Festinger, Leon, A THEORY OF COGNITIVE DISSONANCE, Row, Peterson and Co., 1957, p.3.

17. Waetjen, Walter B., "Learning and Motivation: Implication for the Teaching of Science", READING IN SCIENCE EDUCATION FOR THE SECONDARY SCHOOLS, New York: The MacMillan Co., 1969, p.91.

18. Piaget, Jean, "The Child and Reality: Problems of Genetic Psychology", translated by Arnold Rosin, London: Frederic Muller, 1974.

REFERENCES - CONTINUED

19. Suchman, J.R., "Inquiry Training in the Elementary School", THE SCIENCE TEACHER, Vol. 27, Nov. 1960.

20. Marlins, James G., "A Study of the Effects of Using the Counterintuitive Event in Science Teaching on Subject-matter Achievement and Subjectmatter Retention of Upper-elementary School Students", Doctoral Thesis, The American University, 1973.

21. Liem, Tik L., "A Study of the Effects of Using Discrepant Events in Science Teaching on Concept Retention of Upper Elementary School Students", WORLD TRENDS IN SCIENCE EDUCATION, McFadden, C.P. (Ed), Atlantic Institute of Education, Halifax, 1980, pp. 287-293.

22. Liem, Tik L., "Effects of Using Discrepant Events on Science Concept Retention of Junior High School Students", Paper presented to the National Co-Educators Conference, Winnipeg, October 1980.

23. Best, J.W., RESEARCH IN EDUCATION, Prentice-Hall, New Jersey, 1977.

CONCEPT **SUB-CONCEPT**

ADDITIONAL DISCREPANT EVENTS

SUPPLEMENT TO

EACH OF THE CHAPTERS

IN THE

FIRST AND SECOND EDITION

AIR   EXERTS PRESSURE

THE WEIGHTLESS WATER

Materials: 1. Two identical empty soda-pop bottles & 2 beakers
   2. One small circular piece of transparency (cut from an overhead transparency or any thin clear plastic)
   3. Two small plastic containers/buckets; a toothpick.

Procedure:
   1. Ask a student to imitate exactly what you do, while each of you stand at opposite ends be hind the front table (give the student one of the bottles).
   2. Fill the bottle completely full with water by pouring from the water-filled beaker.
   3. Put your thumb on the opening of the bottle (sneak the circu lar transparency with the small hole between your thumb and the bottle opening) and turn the bottle upside down over the container/bucket.
   4. Slowly let go of your thumb (laughter, because the water will pour out of the student's bottle and not from yours) - See Sketch: Bottle III.
   5. Hold the toothpick vertically and show that you can let it pass the bottle opening and float into the bottle! - See Bottle I & Bottle II.

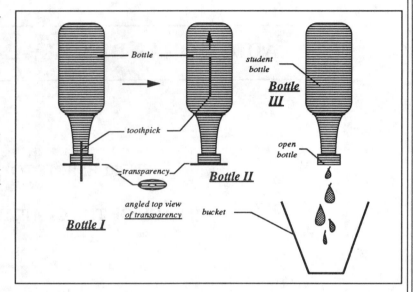

Questions:
   1. Why did the water <u>not</u> pour out from bottle II?
   2. What made the water stay up in bottle II, even with the hole in the plastic?
   3. How large a hole can you make in the transparency before the water will start to pour out?
   4. What other materials can be used in place of the toothpick and in place of the transparency?
   5. What other liquids will behave the same way?

Explanation:
   Just like in Event 1.7 (ISI-2nd Ed), it was air pressure that held up the water in the upside down bottle. Even with the hole in the transparency, the water does not flow out because of surface tension - cohesion of water to water molecules at the surface of the water. Too large a hole in the transparency will make the water pour out. Any small object with a diameter smaller than the hole in the plastic and lighter than water may be floated in the water.
   A transparency is used so that people will not detect it from a distance of about 10 feet or more. If it is not important to conceal this disc, it can be replaced with one made from a 3x5 card or any other stiff sheet.
   If a transparency is not available, a piece of gauze or cheese cloth may be used in its place. One of the holes in the cloth can be made a little larger, such that the toothpick can fit through the hole, and the same effect can be obtained. The reason for the water not pouring out of the hole is the **surface tension** of the water. The molecules of water at the surface pull at each other forming a film, which is pushed up by the **atmospheric pressure**. This **surface tension** can be broken by touching the surface, shaking the bottle, holding the bottle side ways, or by adding a drop of detergent to the surface.

AIR EXERTS PRESSURE
EXPANDS WHEN HEATED

THE BALLOON IN THE JAR

Materials: 1. A wide mouth jar (pickle jar) plus lid.   2. A round balloon.
3. Two deep steel pans, some table salt.   4. An electric or gas stove.

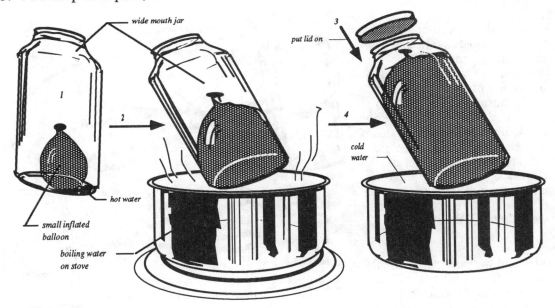

Procedure:
1. Inflate the balloon slightly by blowing (just large enough to slide it into the jar) and place a knot in it, then put the balloon into the jar plus a little water. (see Sketch **Step 1**)
2. Fill both steel pans with water and place one of them on the electric or gas stove pit, put one table spoon of salt in the water and bring it to a boil.
3. Place the jar with the balloon in it in the boiling water (hold the jar with kitchen gloves) and wait till the water inside the jar is steaming hot. (see Sketch **Step 2**)
4. As soon as you see lots of water vapor coming out of the jar, screw the lid on tightly (**Step 3**).
5. Place the jar in cold water and roll it quickly around so that it is cooled off evenly on all sides (see Sketch **Step 4**). Observe the balloon inside the jar!

Questions:
1. What was the purpose of the water inside the jar?
2. What was the purpose of the salt in the boiling water?
3. What made the balloon expand after putting the jar in the boiling water (Step 2)?
4. What made the balloon expand after putting the jar in the cold water (Step 4)?
5. Would the balloon keep expanding the same way if Step 3 (putting lid on) was left out? Why? Why not?
6. What would happen to the balloon if no water was added to the jar in the beginning?
7. In what ways can we reduce the size of the balloon after it is inflated in the jar?
8. What application in daily life can you mention that uses the same principle?

Explanation: This particular demonstration may be used to evaluate students' understanding of the concepts of *AIR EXERTS PRESSURE & AIR EXPANDS WHEN HEATED* (after having done Events 1.10 through 1.14 of ISI-2nd Ed). If the student can answer all the questions, you may be assured that he/she has a good grasp of the concepts mentioned.

The water in the jar was needed to be turned into steam when it was immersed into the hot water. This steam pushed almost all the air around the balloon out of the jar. Closing the lid immediately after steam has come out of the jar trapped the air out. Then when the jar was immersed in the cold water, the steam around the balloon **condensed** back into water leaving almost no pressure outside the balloon. This caused the balloon to inflate inside the jar! Would this same event happen if the balloon was not partially inflated before placing in the jar?

Invitations to Science Inquiry – Supplement

AIR   EXERTING PRESSURE

THE CRUSHING POP CAN

Materials: 1. An empty soda pop can.   2. A clear bucket of cold water.
1. An alcohol or Bunsen burner.   4. A pair of tongs.

Procedure:
1. Put a small amount of water in the soda pop can (enough to cover the bottom of it).
2. Hold the can with the pair of tongs and hold it in the flame of the burner (see Sketch A).
3. Keep holding the can above the flame until lots of vapor/steam is seen coming out of the can.
4. Keep the clear container of cold water close by the burner, and plunge the pop can upside down in the bucket of water (see Sketch B). Observe!!

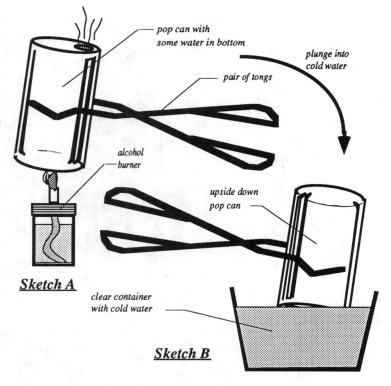

Sketch A

Sketch B

Questions:
1. What did the pop can suddenly do after turning it upside down in the water?
2. What was the function of the water inside the pop can in the beginning?
3. What did the water do when we heated the can above the flame?
4. Why did we have to wait till lots of steam was seen before plunging the can in the water?
5. What would happen if we did not wait till the water was boiling vigorously?
6. What would happen if we plunged the pop can right side up into the water?
7. What is the main cause for the pop can to be crushed?
8. Could we have done this demonstration on the surface of the moon? What things would go differently on the surface of the moon?

Explanation: This domonstration is equivalent to THE COLLAPSING CAN (ISI-2nd Ed Event 1.12). It can be used either for a reinforcing student activity or an assessment of student under standing of the concept of *air exerting pressure*.

The main cause for the pop can to collapse is that the air around the can exerted pressure, and inside the can there was very little pressure left, since the water vapor pushed almost all of the air out (this is why we had to wait for seeing a lot of vapor/steam escaping from the can). At the moment that the can was plunged upside down in the cold water, the water vapor suddenly condensed and left a partial vacuum inside. Since the can was plunged in the water upside down, the opening was closed off by the water. Some water was pushed into the can, but this cooled the can even more down, which caused the water vapor to condense even quicker.

If the can was plunged into the cold water right side up, probably nothing spectacular would have happened. The cooling would have condensed the water vapor, leaving a smaller pressure inside the can, but then the air would have a way to come in and nothing would have happened. If we had a way of closing off the opening before plunging it into the water, it would have crushed the same way. On the surface of the moon: 1. the flame would not be sustained, 2. if we used electric heating, the water would boil at very low temperatures, 3. when plunging the can upside down in the water, nothing would have happened.

AIR

TECHNOLOGICAL APPLICATION
TRANSFERABILITY OF PRESSURE

TURN A LITTLE WATER INTO A LOT OF LEMONADE

Materials: 1. Two identical gallon (4 litre) tin cans. 2. Two 2-hole stoppers (to fit the opening of the cans). 3. Glass tubing, rubber tubing, and a glass funnel.

Procedure:
1. Fit the glass and rubber tubing, and the glass funnel in the two hole stoppers as shown in the sketch above (See Sketch A).   2. Fill Can A about 3/4 full of water (add red coloring for lemonade effect), and pour about 100ml of water in Can B.
3. Make sure that the stoppers are pressed tightly in the can openings. Now you are ready for the demonstration (do not reveal the construction of the tubes inside the cans - students only see things as in Sketch B).
4. Pour some water in the funnel and say: "I am turning a little water into a lot of wine/lemonade," "Can you find out what the structure of the glass tubes inside the cans have to be to make it work like this?"
5. Have students work in groups, and have each group come up with their hypothesis and explanation.

Questions: *(Before seeing the inside structure)*
1. Did you observe any water flowing from Can B to Can A?
2. What would cause the water to flow from Can A?
3. What would adding water to the funnel do to the water level in Can B?
4. Do either of the two tubes in Can B have to extend into the water level or not, or both?
5. The same question (4) for Can A?
(After seeing the correct inside structure):
6. What would happen if the funnel did not extend into the water?
7. What would happen if the glass tube in Can B also extended into the water?
8. What would happen if the long tube in Can A did not extend into the water?
9. What would happen if both tubes in Can A extended into the water?

Explanation:
With the correct structure in *Sketch A* above, after pouring some water into the funnel, the water level in **Can B** rises, increasing the pressure of the air above the water, thus also increasing the air pressure in **Can A**. This will push down the water level and thus push the water up in the long tube. The water drips into the funnel and the cycle is repeated again. The flow of water will stop as soon as either the water level in **Can A** gets lower than the opening of the long tube, or the water level in **Can B** reaches the opening of the short tube.

AIR   AERODYNAMICS

MAKE AN AIR BULLET SHOOTER

Materials: 1. A medium sized cardboard box.
2. A heavy polyethylene sheet (to cover one end of the box).
3. Four long heavy rubber bands, a rubber stopper.
4. Two small petrie dishes, conc. ammonia, conc. hydrochloric acid.

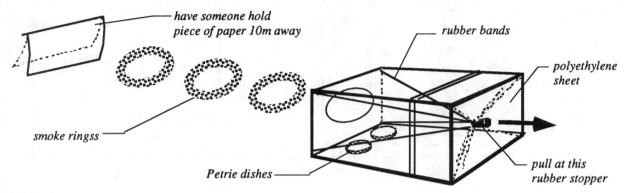

Procedure:
1. Cut a circular hole of about 15 cm diameter in the bottom of the box, and flip the flaps of the open side inwards.
2. Staple the long rubber bands to the four corners of the side with the hole, and staple the other end of the bands to the rubber stopper.
3. Cover the open end of the box rather loosely with the polyethylene sheet and tape the edge of the sheet tightly against the box.
4. Push the rubber stopper against the center of the sheet, wrap the sheet around the stopper, and tie it with a small thick rubber band.
5. Now the air shooter is ready. Have someone hold a paper or cloth sheet about 10 meters away and shoot air bullets at the sheet by holding the box, aiming the circular hole at the sheet, pulling the rubber stopper and release it suddenly (see Sketch above).
6. The air shooter can also be used to produce smoke rings. Tape the two small Petri dishes next to each other inside the box, place a few drops of conc. ammonia in one and a few drops of conc. HCl in the other dish, and shoot smoke rings!

Questions:
1. What makes it possible to blow at the paper sheet from 10 m away?
2. Would a square or triangular opening have the same effect?
3. What shape would the "air bullet" have?
4. What did the smoke consist of?

Explanation:
The circular opening in the box made it possible for the air to form ring shaped"bullets"that can move through the stationary air much faster and farther. Other shapes of the opening in the box would hamper this natural travel of fluids. The smoke was formed by the gases or vapours of ammonia and hydrochloric acid: $NH_3 + HCl \longrightarrow NH_4Cl$, forming ammonium chloride, which is a solid (smoke consists of finely devided solid particles in a gas).

Fill the air shooter box with air freshner mist or perfume, and shoot perfume "bullets" to the people in the audience!

AIR   AERODYNAMICS

THE SMOKE RING RACE

Materials: 1. The air bullet shooter (see Event 1.42 ISI-2nd Ed)
  2. Cigarettes (if you are able to make/blow smoke rings).

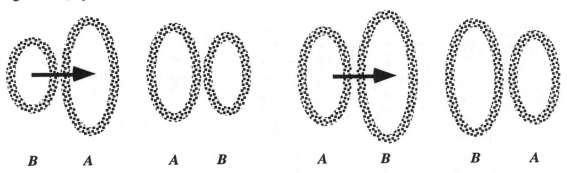

B   A   A   B   A   B   B   A

Procedure:
  1. Place a few drops of conc. ammonia in one petri dish, and a few drops of conc. hydrochloric acid in the other dish in the air shooter box (see AIR BULLET SHOOTER-page17).
  2. Make a slow moving smoke ring by pulling the stopper halfway or not as far out of the box and releasing it.
  3. Immediately after making this slow moving ring, produce a fast moving smoke ring by pulling the stopper out much farther and releasing it. When making this second ring, do not change the direction of the box opening, in other words, shoot directly at the first smoke ring.
  4. Observe what the rings are doing!
  5. If you are able to blow cigarette smoke rings: make a large ring, then blow a smaller ring through this first one by holding your mouth opening a little narrower. (If the rings made with the air shooter are of the same size, hold a cardboard with a smaller circular opening in front of the air shooter to make the second smaller smoke ring).

Questions:
  1. Do you notice that the rings are chasing each other?
  2. What did the first ring do when the second ring went through it?
  3. Which of the rings actually won the race?
  4. Did the rings stay the same size during the race?
  5. Which size rings would travel faster? The smaller or the larger ones?
  6. Why would a ring be formed in the first place?

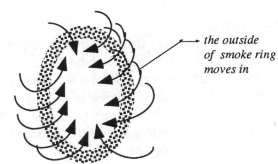

the outside of smoke ring moves in

Explanation:
  The circular opening in the box (or the mouth) makes the air in the center of the circle move fastest. The ouside of the circle of air is held back somewhat, thus a circular motion is created, causing a ring to be formed. The smaller second ring will flow with the air in a circular path and thus become larger as soon as it has gone through the first ring, while the first ring gets smaller and passes through the second, etc.

FLOWING AIR

WIND ENERGY
BERNOULLI'S PRINCIPLE

THE LEAPING EGG

Materials: 1. Two identical wine glasses (or plastic cups).
2. A hard-boiled egg (or ping-pong ball).

Procedure:
1. Place the two wine glasses or cups about 2-3 cm apart on the table and secure them down with tape or just hold them down.

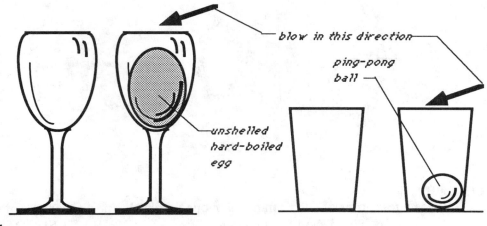

2. Put the hard-boiled egg in one of the glasses (or the ping-pong ball in one of the cups) and ask the audience the question: "How can I move the egg (or ball) from one glass into the other without touching the egg and leaving the glasses as they are?"
3. Most people will respond with: "It's impossible!" Now blow a short and hard puff obliquely into the far side of the wine glass that holds the egg or ball and watch the egg leap! (It may take a few practice blows to make the egg leap successfully).

Questions:
1. What makes the egg or ball jump out of the first glass?
2. Why would the egg or ball not go toward the one that blows?
3. What does flowing air create?
4. How far apart can we place the second glass in order for this one still to catch the leaping egg or ball?
5. What would happen if we blew on the near side of the egg or ball?

Explanation:
Blowing obliquely into the far side of the glass builds the pressure on that side. It pushes the egg or ball out of the glass and the flowing air above both glasses guides the egg towards the second glass, because the flowing air actually creates a lower pressure. The harder we blow, the farther the second glass can be placed to catch the leaping egg. (Warning: <u>do not</u> try this activity with a raw egg!)

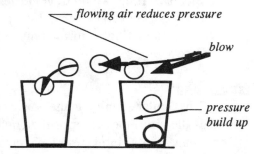

Again, this would be an application of ***Bernoulli's Principle***, which states: ***the faster the flow of a fluid, the lower the pressure.***

$$P + \frac{KE}{V} = \text{Constant}$$, where P=pressure, KE=kinetic energy of fluid and V=volume of fluid.

Invitations to Science Inquiry – Supplement – Page 30

FLOWING AIR
FLOWING FLUID
BERNOULLI'S PRINCIPLE

LIFT A MOTH BALL WITH STREAMING WATER

Materials: 1. A small plastic vial (medicine vial or narrow jar).
2. A moth ball (or raw egg plus a small drink glass).
3. A water tap above a sink.

Procedure
1. Place the moth ball in the vial (or egg in the small glass), then ask the students: "How can I lift the moth ball (or egg) up without turning the vial or glass upside down?"
2. Now open the tap and let the water run with a smooth medium size flow (this should be tried out beforehand so that you know what size stream you need for the lifting power).
3. Hold the vial with the moth ball or glass with the egg under the water stream, such that the water falls directly on top of the ball (or egg). Observe the ball (or egg) rise to the water surface!

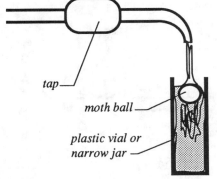

Questions:
1. What made the moth ball rise up in the water?
2. Does a moth ball sink or float in water?
3. What happens if the water flow is suddenly stopped?
4. Would the moth ball keep floating if the water keeps running?
5. What would happen if the water stream was not falling head on?
6. What would happen to the ball if we used a larger/wider jar?
7. What happens if we used a stronger or weaker flow of water?
8. After the ball got lifted to the top, what would happen if the water stream were increased or decreased?
9. What happens to the ball if the vial was held in a slanted position?

Explanation:

Bernoulli's Principle states: *the faster the fluid flow, the lower the pressure.* In this case the fluid is the water streaming out of the tap. The faster this flow of water, the lower the pressure, but the larger the downward force. This means that there is an optimum flow where it creates a low enough pressure above the ball. This makes the ball rise up with the water flow until it reaches the surface.

A sudden stop of the water flow will make the moth ball sink, showing that indeed the lower pressure created by the water flow was the cause for the ball going up. Increasing the water flow will also make the ball sink, but this time because of the downward force of the water.

A technological application is the Venturi tube, a measuring instrument for fluid speed based on this same Bernoulli's Principle (see Sketch on right). The difference in pressure between the smaller tube (P2) and the larger tube (P1) will indicate the speed of fluid flowing through the tube.

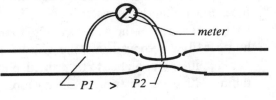

FLOWING AIR   FLOWING FLUID
                                                 APPLICATIONS

THE SELF-PRIMING SIPHON

Materials: 1. A short wide glass tube, open on both ends (diam 2.5 cm, length about 8-10 cm long).
2. A medicine dropper (or glass tube drawn at one end).
3. One-hole stopper with a short glass tube.
4. Two-hole stopper (both stoppers fitting in the wide tube).

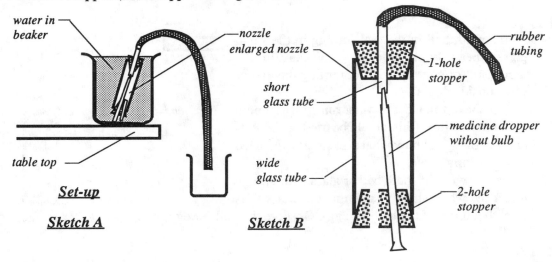

Procedure:
1. Insert a short glass tube into the one-hole stopper and connect a long rubber tubing to the glass tube.
2. Take the rubber off from the medicine dropper and insert the glass tube into one of the two holes of the two-hole stopper.
3. Fit both stoppers into the wide glass tube, which is open on both ends, and make sure that the drawn end of the medicine dropper extends into the opening of the short glass tube which is inserted into the 1-hole stopper (see Sketch B).
4. Plunge the wide tube with stoppers and all in a large water container on the table and bring the end of the rubber tubing below the table top. (the extent to which the drawn tube goes into the top glass tube may need some adjusting).

Questions:
1. What is the cause for this siphon to start by itself?
2. What happens to the air inside the wide tube at the moment that it is plunged into the water?
3. What does the drawn tube create when water is pushed through it?
4. How is a regular siphon started?

Explanation: At the moment that the wide tube is immersed into the water, water comes up in the wide tube through the open hole in the 2-hole stopper. This causes the air inside the wide tube to flow through the small tube in the 1-hole stopper. Simultaneously, water is coming up through the drawn tube, and a small jet of water is created. Both flows of air and water create a lower pressure and thus an upward surge of the water, enough to overcome and fill the hump of the rubber tubing. Once this hump is filled with water and the wide tube is also filled with water, the siphon will continue running.

A regular siphon has to be primed, either by sucking through the rubber tubing, or by filling the tubing with water - by immersing the tubing in the water, pinching the end and pulling it quickly over the edge of the container and letting go of the tubing.

FLOWING AIR   RESISTANCE

THE SELF-DIRECTING CARDS

Materials: 1. A deck of playing/bridge cards

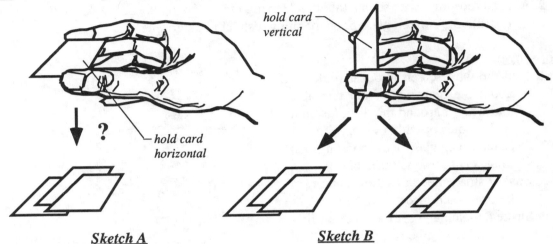

Sketch A   Sketch B

Procedure:
1. Stand up straight with a deck of cards in your hand. Hold one card at a time in front of you between two fingers horizontally and ask: "Where do you think this card will end up on the floor? Straight underneath it or off to the side?", then drop the card. (See Sketch A).
2. Hold the next card between thumb and index finger by the short sides, vertically with a slight slant to the left; ask the same question, then drop the card (See Sketch B).
3. Hold the next card the same way, but now slant it slightly to the right, ask: "Where will this card end up on the floor?", then drop the card.

Questions:
1. Where did the first, second and third card end up on the floor?
2. What made them fall so differently?
3. Which of the cards stayed steady while falling?
4. How could you separate the cards into three equal piles by dropping them to the floor? (This question could be used before doing the demo.).
5. What do you think would happen if you held the cards by the long sides and let them fall from the three different positions?
6. What do you think the cards would do when dropping two at a time?
7. Where would the cards end up on the floor, if the cards were held in a random fashion before dropping?

Explanation:
When letting the card fall horizontally, it will stay horizontal and end up quite plumb underneath the position before falling. This is caused by the even flow of the air around the card. When holding the card vertically, the leading edge of the card cuts into the air, encounters a slightly higher pressure or resistance on one side, flips up, and the card starts rotating. Once it spins it will keep on spinning. The ones slanted to the left will rotate in a clockwise direction, and the ones slanted to the right will rotate counter-clockwise (from the point of view of the performer).

This demonstration must be done in a draft-free room. It will not have the same results outdoors. On a windy day, dropping the cards will have the same random result as holding the cards at random positions indoors before dropping them.

WEATHER   CYCLONE

MAKE A FIRE CYCLONE

Materials: 1. An old turntable or "lazy susan" (or a circular wooden board on marbles).
2. A metal mosquito wire screen (about 30 cm high to fit around the turntable (or the turning wood board). 3. A watch glass or evaporating dish, cloth, kerosene.

Procedure:
1. Roll the wire screen into a cylinder of about the diameter of the turntable and wrap it around the disk, fasten it with staples or with a wire.
2. Place the watch glass or evaporating dish in the center of the turntable.
3. Soak a small piece of cloth (cotton) in kerosene and place it on the dish.
4. Strike a match and set the cloth on fire, wait till the flame is a good size, then switch the turntable on (or start rotating the circular board).
5. Observe the flame shape!

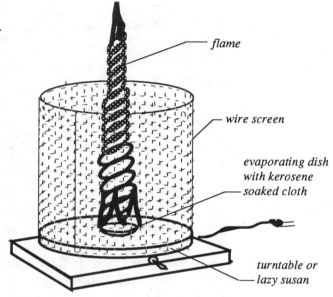

Questions:
1. What hypotheses/inferences can you make to explain the event?
2. What does a flame need in order to sustain burning?
3. Where does the air come from to reach the flame?
4. In which direction do the air molecules move with respect to the rotation?
5. What shape did the fire take shortly after the rotation started?
6. What could we use instead of kerosene?
7. What event in nature does this resemble?
8. What is the reason for the flame to take that particular shape?

Explanation:
   The reason for the flame to be shaped like a *cyclone* is the direction from which the air feeding the flame has to come. Because of the rotating screen around the flame, the air molecules are given an initial **angular momentum** at the moment it enters the screen cylinder from outside. This angular or **rotational momentum** shapes the flame into this cyclone.
   In nature, cyclones are formed when two winds of opposing direction pass each other. At the border of the two passing masses of air, a *cyclone or vortex* can be formed. These cyclones are also called tornados or hurricanes. They are destructive wind storms that can occur over a large area, like hundreds of miles wide. The winds forming the vortex blow spirally inward, creating a low barometric pressure in its center. In the Northern hemisphere these cyclones rotate counter-clockwise and in the Southern hemisphere in a clockwise direction.
   These vortexes are also formed in water under waterfalls, but they are not going up but are cone-shaped and downward sucking. It is therefore quite dangerous to swim in lakes close to a waterfall.

Invitations to Science Inquiry – Supplement – Page 34

CHARACTERISTICS OF MATTER **MOLECULAR NATURE**

THE COOLING RUBBER BAND

Materials: 1. A wide rubber band, match and ruler. 2. A stand and a 100 gram weight.

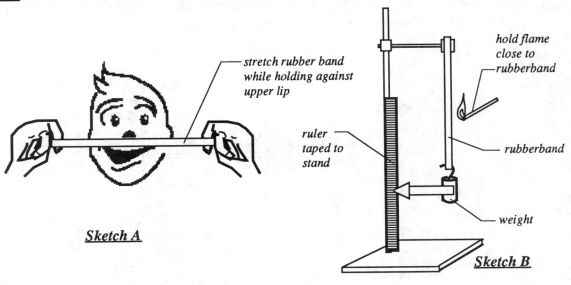

Sketch A

Sketch B

Procedure:
1. Hold the rubber band partly stretched between your forefingers, such that the band touches your upper lip.
2. While the rubber band touches your lip, stretch the band fully and release the tension slowly. What do you feel? (see Sketch A).
3. Hang the rubber band from a stand and hook a 100 gram weight to the band (tape a paper arrow to the weight pointing to the vertical ruler (see Sketch B).
4. Note the position of the paper arrow.
5. Now strike a match and hold the flame close to the rubber band. Observe the position (reading) of the paper arrow.

Questions:
1. How did the rubber band feel (against the upper lip) when it was stretched? When the tension was released?
2. What would stretching of the rubber do to the rubber molecules?
3. What did the rubber band do in demonstration B, when the flame was held near to it?
4. What do metals and other materials generally do when heated?
5. What other material behave the same way as rubber when heated?

Explanation:
Rubber is a high polymer, which means that it consists of large molecules with many side branches. When stretched, these molecules bump and slide against each other, creating heat. This makes the rubber band feel warm against the upper lip. When the tension is released, the molecules can relax back into their normal positions and cooling takes place in the rubber. This cooling process in the rubber means that it is withdrawing heat from its environment.

In demonstration B, this heat is supplied by the environment and forces the rubberband to contract. Other materials that are composed of polymers behave similarly, like: polyethylene, polystyrene, and other plastics.

Invitations to Science Inquiry – Supplement

CHARACTERISTICS OF MATTER   STATES OF MATTER

GET THE THREE STATES OF MATTER OUT OF WOOD!

Materials: 1. Alcohol or Bunsen burner.   2. Two wide test tubes & 1-hole and 2-hole stoppers.
3. Two bent glass tubings (fitting in the stoppers).   4. A small beaker/cup, wood splints.

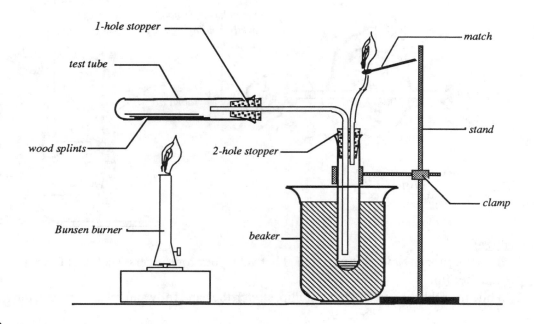

Procedure:
1. Place a few (5-6) wood splints in one of the wide test tubes (the horizontal one) - see sketch.
2. Set the equipment up as shown on the sketch.
3. Start heating the horizontal test tube (with the wood splints) with the alcohol or Bunsen burner.
4. Test the produced gas, ejected from the short bent tube, for combustibility by holding a lit match over the opening.
5. Keep heating the wood splints until they all turn black.

Questions:
1. What were the three states of matter produced?
2. What was the solid substance left in the horizontal tube?
3. What was the gas that was ejected? Was it flammable?
4. What liquid was condensed out in the vertical tube?
5. What is the function of the water in the beaker?
6. Can the three substances be combined to produce the wood?

Explanation:
The wood splints in the horizontal tube consists of cellulose, which breaks down by heating in the absence of air (oxygen) - *dry distillation* - producing a mixture of gases: methane, carbon monoxide, carbon dioxide, water vapor, etc., and a mixture of liquids: acetaldehyde, vinegar, acetone, and a mixture of alcohols and tars. The black remaining solid in the horizontal test tube is charcoal, which may be burned further to ashes in the presence of air (oxygen). This process of heating a substance in the absence of air (oxygen) is called *dry distillation* and is utilized in the production of charcoal from wood or coke from coal.

CHARACTERISTICS OF MATTER

COLLOIDS
STATES OF MATTER

MAKE MILK FROM WATER AND OIL

Materials: 1. A glass/plastic clear jar (or beaker), stirrer.   2. Cooking oil, liquid detergent.

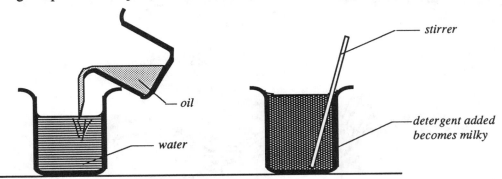

Procedure:
1. Fill the jar about half full with water and pour half of that volume (1/4 of the jar volume) of oil over the water.
2. Stir the two liquids with a spoon or glass stirrer and leave it for a while, and observe what is happening to the mixture.
3. Ask the students: "How can I make the two liquids stay mixed?" Now add a few squirts of liquid detergent and stir thoroughly.
4. After mixing thoroughly, leave the jar alone and observe (if the emulsion separates again you may need more detergent or more vigorous stirring or both).

Questions:
1. After stirring without detergent, what did you observe in the jar?
2. What made the two liquids stay mixed as an emulsion?
3. How would you define "emulsion"?
4. What is the term for a finely divided solid in a liquid?
5. Can you name some examples of finely divided solids in a liquid?
6. What is a finely divided solid in a gas? Examples?
7. What do we usually call fine droplets of liquid in the air?
8. What do we call finely divided gas bubbles in liquid?

Explanation:
When mixing oil and water, the oil will break up in small droplets and be dispersed in the water very temporarily. After leaving the jar with the mixture alone for a while, the two clear liquids will separate: oil forming a layer above the water because of its smaller density. After adding some liquid detergent (the *emulsifier*) and some vigorous stirring, the small droplets of oil will stay dispersed forming an *emulsion*. Examples of an emulsion: milk, mayonnaise, salad dressings, butter, etc.

Finely divided droplets of liquid in gas is mist or fog, colloidal solid in liquid is called a *suspension*: muddy water; a solid *colloid* in gas is called smoke. An example of a colloidal dispersion of solid in solid is gold particles in ruby glass; gas bubbles in liquid is commonly called: *foam*. Examples of foam are: the regular foam in detergent solutions in water, foam in beer. The "foam rubber" is actually dispersion of gas in a solid, so is "styrofoam".

Invitations to Science Inquiry – Supplement – Page 37

CHARACTERISTICS OF MATTER

STEAM & MOISTURE
STATES OF MATTER

THE INVISIBLE STEAM

Materials: 1. An ordinary steam kettle (for boiling water). 2. Large candle or an alcohol burner.

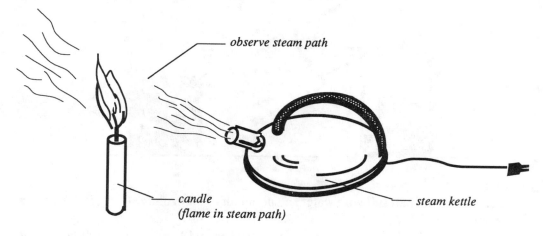

Procecedure:
1. Boil some water in the kettle.
2. As soon as the water boils, observe the steam leaving the kettle spout.
3. Light the candle/alcohol burner and place the flame slightly under the path of the steam. What do you observe?
4. Move the burning candle/alcohol burner to another spot in the steam path; what do you observe about the steam?

Questions:
1. Did you observe any steam above the burning candle flame?
2. What makes steam visible to the human eye?
3. How is fog formed in the atmosphere?
4. Does cold or warm air contain more moisture?
5. Why are clouds mostly formed at higher elevations?
6. What are conditions for fog to form?

Explanation:
   The steam is not visible in hot or warm air, because steam can only be seen when it is in the form of tiny *condensed droplets.* These droplets turn into vapor at higher temperatures. Fog and clouds are only formed when the atmosphere contains a lot of moisture and when the temperature is so low that the water vapor condenses. These lower temperatures are present usually at higher elevations.
   Moisture comes in the atmosphere from the soil, from plants, even from the breath that humans and animals exhale, but mostly from the ocean. Try exhaling against a mirror and you will see condensation of vapor.
   Hot air can contain much more water vapor than cold air. When the atmosphere is saturated with moisture, the decreasing of the temperature will precipitate out excess moisture. This precipitate is in the form of fog or rain or snow, depending on how low the temperature is.

CHARACTERISTICS OF MATTER — MOLECULAR SPACING

IS THE CUP REALLY FULL?

Materials: 1. Two clear plastic cups (or drinking glasses)
2. Absorbent cotton balls (or new Pamper diapers)

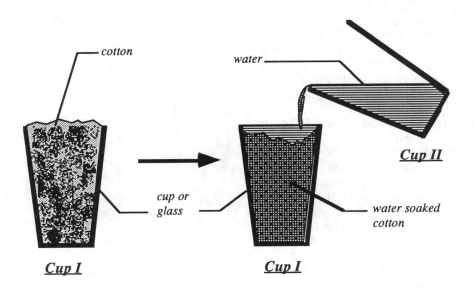

Procedure:
1. Fill one of the cups full with cotton balls (or the inside of the Pamper diapers) and fill the other cup almost full with water.
2. Ask the students: "Do you think I can pour all this water in the cotton-filled cup without letting it overflow (anticipated answer: "No way!")
3. Then proceed to pour <u>all</u> the water slowly into cup No.1 without letting one drop overflow!

Questions:
1. Why did the water not overflow?
2. How much space did the cotton actually take in the cup?
3. What was the shape of the water meniscus in cup I?
4. How much additional water can be added to the full cup?
5. Would this same demonstration work with oil? Alcohol? Other liquids? Non-absorbent cotton?

Explanation:
When compressed, the cotton would most likely take very little space. Besides, the cotton is *water absorbent,* which means that the water molecules can come very close to the cotton molecules, and thus they can slip into each other. Because of these characteristics of the cotton and the *space that exists between the water molecules*, the water can **all** be poured into Cup I without overflowing. The material inside a Pamper diaper is **polyacrylate**, and is so absorbent that the water is actually bound by the chemical. The mixture turns into a gel and the water cannot be poured out anymore.

This polyacrylate powder is so water absorbent that it can hold 400 times its weight in water. An intriguing demonstration is to pour water in an opaque cup holding some polyacrylate powder (which will turn into a gel), and by turning this cup upside down no water pours out! The water disappeared!

CHARACTERISTICS OF MATTER

SINKING & FLOATING COMPRESSABILITY

THE MAGNETIC FINGER

Materials:
1. An empty 2-liter soda pop plastic clear bottle.
2. A small medicine dropper
3. A tall drinking glass.

Procedure: *PREPARATION*
1. Take the stiff bottom of the clear, plastic bottle off by forcing it down from the transparent part and then completely fill the bottle with water. (Use the stiff part as a stand on the table)
2. Fill the tall drinking glass full with water.
3. Test the dropper for boyancy: fill it half full with water by pushing the rubber bulb and sucking water into it; place it in the drinking glass - the dropper should **just** float. (If it sinks, push a few drops out of the dropper and test again).
4. Now place the dropper in the water-filled bottle and cap the bottle quite tightly. Now you are ready for the demonstration.

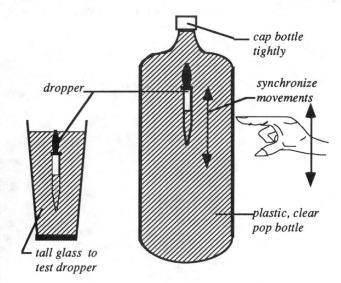

DEMONSTRATION
5. Hold the bottle by the bottom in your left palm and point with your right index finger to the dropper, saying: *"As soon as I touch the bottle, my finger will become a magnet and the dropper comes with it"*. - Move your right finger down along side the bottle and simultaneously press the bottle with your left hand.
6. Practice the synchronization of finger movement and pressing!

Questions:
1. Does your finger really become magnetic?
2. What makes the dropper sink?
3. What did the water level inside the dropper do when the left hand pressed the bottom of the bottle?
4. What did we have to do to make the dropper move up?
5. What is more compressable? Water or air?
6. Where do we find this principle applied in daily life?

Explanation: By pressing the bottom of the bottle with our left hand (if we are right-handed) and moving our right index-finger down along side the bottle, then easing the pressure and moving the finger up, it seems like the dropper is being pulled down and up by the finger (as a magnet). It is just that the pressing of the bottle and the movement of the finger has to be as synchronized as possible. The pressing of the bottle compresses the air in the dropper (*air is more compressable than water*) and more water gets in the dropper making it heavier, and thus the dropper sinks. When releasing the pressure, the relatively compressed air expands and the water is pushed out of the dropper, making it lighter and thus making it float.

This same principle is used in submarines: in order to submerge, they just need to get heavier. They can do that by pumping in water. To surface, they need to get lighter and this they can do by pumping the water out.

CHARACTERISTICS OF MATTER

DENSITY
IMMISCIBILITY

THE CLEARING SPOT

Materials:
1. A glass Petrie dish and a medicine dropper.
2. Alcohol (methyl-, ethyl-, or isopropyl-).
3. Food coloring.

Procedure:

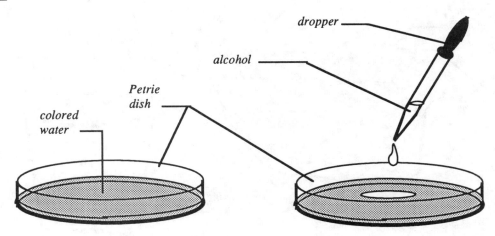

1. Cover the bottom of the Petrie dish with a thin layer (1mm) of water and color with a few drops of food coloring.
2. Mix thoroughly and place the dish on the overhead.
3. Put a few drops of alcohol in the center of the Petrie dish with the medicine dropper. What do you observe?

Questions:
1. What did you observe after the alcohol was added?
2. What would happen if the water layer was much thicker?
3. What other liquids can be used to replace the alcohol and still obtain the same effect?
4. What would happen if we exchanged the liquids? (a few drops of water on colored alcohol?)
5. What would happen if the alcohol was dropped on the side of the dish? (not exactly in the center?)
6. What do you observe if you drop the alcohol close to the water edge in the clearing hole?

Explanation: The alcohol drops will push the thin water layer away and form a small colorless circle in the center of the Petrie dish. This happens because alcohol is rather *immiscible* with water, meaning that the two liquids do not mix when they are not stirred. If the two liquids were *miscible,* no clearing spot would be formed, this is also the case when the water layer is too thick.

After the clearing spot is formed, let a few alcohol drops fall close to the water edge inside the clear spot: the water edge forms a protrusion and wrinkles are formed; the colored edge moves as if it were "swallowing" the alcohol and moves back. This occurs because the alcohol slowly mixes with the water. Over time the clearing spot becomes smaller and smaller until the bottom of the dish is completely covered again. This happens because the alcohol evaporates and thus there is nothing to repel the water.

Try to do this demo with a plastic Petrie dish!

CHARACTERISTICS OF MATTER

KINDLING POINT

THE EXACT PAPER CIRCLE

Materials: 1. A regular drinking glass.   2. A paper napkin and matches.

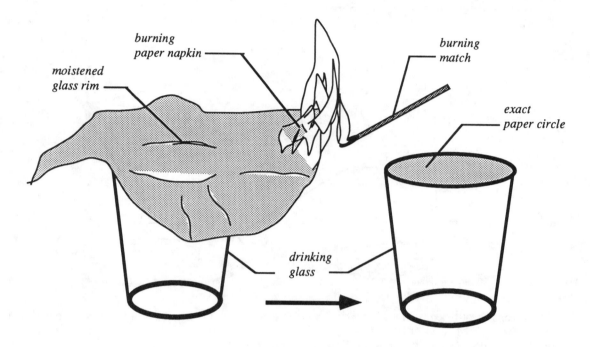

Procedure:
1. Fill the glass half full with water and ask a question: "How can we make a circle out of the napkin that is exactly the size of the glass opening?"
2. Moisten the rim of the glass with some of the water in the glass; tear a piece of napkin slightly larger than the opening of the glass.
3. Hold the paper tightly stretched over the glass and place it on the wet glass rim.
4. Strike the match and burn the napkin. Observe!

Questions:
1. What made the flame stop exactly at the glass rim?
2. How else can we make a paper circle exactly the size of the glass opening?
3. What would happen if we had more than one layer of paper on top of the glass?
4. Can this demonstration be performed with a plastic glass?
5. What is the purpose of moistening the glass rim?
6. What would happen if the paper was not taut but wrinkly on top of the glass opening?
7. What would happen if the paper was too wet at the rim of the glass?

Explanation: The purpose of moistening the glass rim is to wet the paper exactly at the rim. This lowers the kindling point of the paper and thus the flame will stop exactly at the rim. If the paper is not totally wet at the rim, when two or more layers of paper are used, then there is a possibility that the flame will travel beyond the rim. Wrinkly paper also leaves this same possibility. A plastic glass might melt as soon as the heat of the flame reaches it.

    This demonstration shows that water lowers the kindling point of just about anything, and that is why it is used as a general fire extinguisher. To prevent houses from burning during an outbreak of a fire, water is sprayed on adjacent buildings, so that the kindling temperature of these buildings is much lowered.

CHARACTERISTICS OF MATTER   ABSORPTION SURFACE TENSION

THE FUNNY TOOTHPICKS

Materials:
1. A dozen toothpicks (preferably the flat ones, but round is OK too)
2. A dinner plate, or any other flat container for water.
3. One or two sugar cubes. 4. A piece of soap or a drop of liquid detergent.

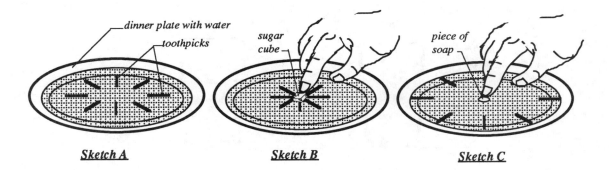

Sketch A   Sketch B   Sketch C

Procedure:
1. Fill the dinner plate almost full with water
2. Let 6-8 toothpicks radially float on top of the water - place them in the middle between the plate's center and the edge. (see Sketch A above).
3. Take a sugar cube and let the cube touch the water in the center of the plate and hold it there. Observe what the toothpicks are doing (see Sketch B above).
4. Now take the piece of soap (or the dropper with liquid detergent) and touch the water with the soap in the center of the plate. What are the toothpicks now doing? (Sketch C)

Questions:
1. What made the toothpicks come together when we toughed the water with the sugar cube?
2. What is the sugar actually doing with the water?
3. In what direction did the water actually flow when the sugar cube touched the center?
4. Why did the toothpicks fly apart when the soap touched the center of the plate?
5. What does soap do to the surface tension of water?
6. What other materials can be used to replace the sugar cube and get the same effect?
7. What substances can be used to replace the soap or detergent?
8. What other materials can be used instead of the toothpicks?

Explanation:
   The sugar cube is quite porous and has many capillary cavities in between the sugar crystals. Therefore when we touch the water surface in te center of the plate with the sugar cube, water is immediately *absorbed* into the cube, because of *capillary action.*. Since the water at the surface flows towards the center, the toothpicks are coming with the flow, which is towards the center. Touching the water with soap in the center of the plate breaks the *surface tension* at this spot, eliminating the adhesive force at one end of the toothpick (the force between the water molecules and the toothpick). The resultant force between water molecules and toothpick is therefore from the center radially outward.
   Other materials that may replace the sugar cube: a sponge, absorbent cotton, small capillary tubes, any water absorbent material. Replacing the soap or detergent: oil or alcohol, which will spread on the water surface and thus push the toothpicks radially out. This, however, is **not the same principle** as the breaking of the surface tension of water! Instead of the toothpick we may use any other material that floats on water: like, wooden matches, small corks, etc.

CHARACTERISTICS OF MATTER

SURFACE TENSION

HOW MANY PENNIES CAN GO IN ?

Materials: 1. Two regular drinking glasses, liquid detergent.
2. About 50 pennies (or other small coins).

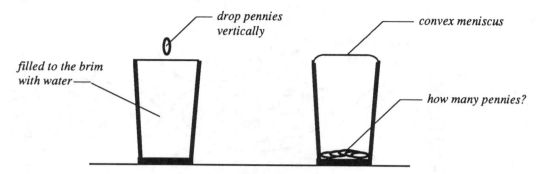

Procedure:
1. Make sure that the drink glasses are very clean. Place one on the table and fill it to the brim (not too overfull, just full).
2. Now ask the students: "How many pennies can I put in the glass before it overflows?" Anticipated answer: "5, 10, 15, even 20 maybe".
3. Start putting the pennies in the glass of water, very carefully with its edge in first (vertically), and let the students count.
4. Surprised at the result? Now place the other glass on the table, in which you put a drop of detergent beforehand (without the students noticing it), and also fill this to the brim with water.
5. Now ask a student to put just as many pennies in this glass. What happened here?

Questions:
1. How many pennies could go in the first glass? In the second?
2. What made the water overflow so easily in the second glass?
3. What kept the water from overflowing in the first glass?
4. What shape did the meniscus take form in the first glass?
5. How would the number of coins compare if we used dimes instead of pennies? Nickels instead of pennies?

Explanation:
It is possible to put close to 50 pennies in the full glass of water, depending on how full the glass was filled. A thick-rimmed glass will look full even when the meniscus is a little below the rim. Then it is possible to put many more pennies in. A larger glass also increases the number of pennies that can be dropped. The water forms a ***convex meniscus*** due to the ***surface tension,*** which is nothing but a manifestation of cohesion of the surface molecules.

In the second glass, where detergent was present, this surface tension is broken. The cohesion between the surface water molecules is much smaller, and thus the water overflows much sooner. (Another way of inconspicuously putting detergent in the second glass, is wetting the second bunch of pennies with a little detergent before giving it to the student).

CHARACTERISTICS OF MATTER

CAPILARITY
COHESION & ADHESION
SURFACE TENSION

THE STRONG SOAP FILM

Materials: 1. A piece of coat hanger or thick wire & thin sewing thread.
2. A medium sized glass or plastic funnel, a candle.
3. Liquid detergent (no additives e.g. JOY), glycerin.

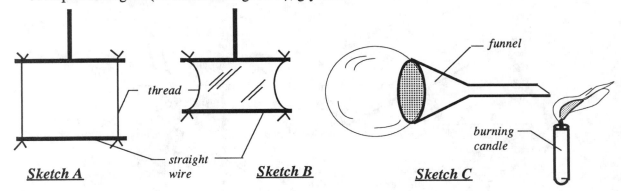

Sketch A Sketch B Sketch C

Procedure:
1. Make a soap solution by mixing 1 part of Joy, 2 parts of glycerin, and 3 parts of water.
2. Tie a short straight piece of wire (about 10 cm) with equal lengths of thread to the T-shaped wire (soldered together beforehand - see Sketch A).
3. Dip the wires into the soap solution and pull them slowly out, holding the T-shaped wire by the leg, and observe (see Sketch B).
4. Light the candle and secure it to the table top.
5. Hold the funnel by its long end and dip it into the soap solution. Blow through the funnel and form a medium sized bubble, but make sure that you let the bubble stick to the funnel.
6. Point the funnel opening to the candle flame and observe the candle flame (see Sketch C).

Questions:
1. What made the threads between the wires curve?
2. What would happen if the soap film between the wires is punctured?
3. What happened to the candle flame?
4. What happened to the soap bubble in the funnel?
5. What do both events indicate about the forces in a soap film?

Explanation:
In the soap film between the wires and thread there exists a force called: *surface tension*. It consists of adhesive forces between the liquid molecules and the solids *(called adhesion),* and cohesive forces between the molecules of the soap solution (called *cohesion).*

This same adhesive/cohesive force or sometimes also called *capillary force* exists in the bubble which is attached to the funnel. This force is pushing the air in the funnel out of the opening and is almost strong enough to blow the candle out.

Examples of capillary forces in action are: the sucking of water from the roots of plants travelling into the top leaves (also called *osmotic pressure);* the glueing of two pieces of plexiglass with a thin organic liquid (by placing a few drops of the liquid between the two solid surfaces).

CHARACTERISTICS OF MATTER

DENSITY
SURFACE TENSION

THE FLOATING OIL SPHERE

Materials: 1. A large, wide test tube or small beaker (100 ml). 2. A medicine dropper.
3. Ethyl or methyl alcohol, frying or light motor oil.

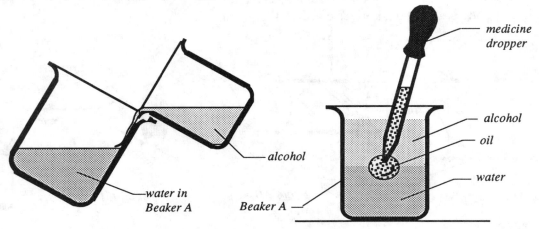

Procedure:
1. Pour about 50 ml of water in the beaker (or test tube).
2. Hold this beaker in a slant and slowly pour the alcohol on top of the water (about 50 ml - an equal amount as the water).
3. Let the beaker stand on the table.
4. Fill the medicine dropper with oil, bring the opening of the dropper in the beaker where the two liquids meet, and squeeze the oil out. Then withdraw the dropper carefully out of the liquid.

Questions:
1. Why does the alcohol float on top of the water?
2. What would happen if we poured the alcohol carelessly on top of the water?
3. Is alcohol completely immiscible with water?
4. Why does the oil take the shape of a sphere?
5. Why does the oil sphere stay between the two layers of liquid?
6. What would happen if the oil was squirted out of the dropper in the alcohol layer and not where the two liquids meet?
7. What would happen if the oil was squirted out in the water layer?
8. What would happen if the oil was squirted out in water only?

Explanation: As ethyl alcohol has a density of 0.794 and methyl alcohol even less than that, it can float on water and form a layer. It is not totally immiscible with water, but when poured slowly and carefully on the water it can form a layer and stay above it. Where the two liquids meet, however, the water and alcohol mix and form a liquid with a **density very close to that of oil.** This is why the oil forms a perfect sphere between the two liquid layers.

   A sphere is formed because it has the smallest surface area as compared to other three dimensional shapes. When the beaker is left standing, the alcohol evaporates slowly, and the oil sphere moves up slowly until it reaches the surface and then the sphere is knotted off at the top and slowly becomes a flat circle when all the alcohol has evaporated.

CHARACTERISTICS OF MATTER — DENSITY

WHICH WILL FLOAT WHERE?

Materials:
1. Mercury, Carbontetrachloride, Kerosene (or gasoline).
2. A piece of steel (bolt or nut), a piece of ebony wood, paraffin or candle), and a cork.
3. A long thin measuring cylinder (100 ml), three small beakers.

Procedure:
1. Pour about 20 ml of mercury in the cylinder, followed by 20 ml each of carbon-tetrachloride, water and gasoline. Do this pouring by holding the cylinder slanted so the liquids will slide on top of each other (Sketch A).
2. Now pick up the piece of ebony, and ask the students: "Where will this object end up, between which layers, or maybe on the bottom?"
3. Pick up the steel bolt and ask the same question (record students answers) then drop it in the cylinder and observe.
4. Do the same with the piece of candle and the cork.

Questions:
1. Where did the piece of steel end up, was it on the bottom?
2. Where did the other objects end up floating?
3. How do the densities of steel and mercury compare?
4. How do the densities of the four liquids compare?
5. If we know that the density of water equals 1, what can we say about the densities of ebony, steel, carbontetrachloride and mercury?
6. What can we say about the densities of paraffin, cork, and gasoline?
7. On what property of the material does it depend, whether an object will float on a liquid or sink in it?

Explanation:

In ascending order the densities of the solids and liquids are as follows: cork, kerosene/gasoline, paraffin/candle, water (1), ebony, carbontetrachloride, steel, mercury. This demonstration shows that even steel can float on a liquid, as long as the liquid has a higher density than steel (see the Table of Densities in the Appendix of ISI - 2nd Ed).

The four layers of liquid can be used to roughly determine the density of an object. This can be done by just dropping the object in the cylinder and observing where the object ends up floating. In the order above, all those preceding water have a density of less than 1. Those that come after water have a density of greater than 1.

CHARACTERISTICS OF MATTER

DENSITY

THE UNLEVEL COMMUNICATING TUBE

Materials: 1. A communicating tube (U-tube or tubular vase) 2. Water and alcohol (methyl or ethyl).
3. Food coloring (green, red or blue) 4. Two small beakers or plastic cups.

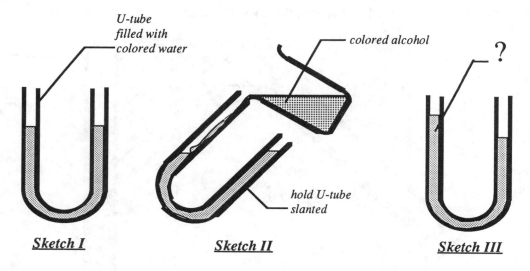

Sketch I Sketch II Sketch III

Procedure:
1. Fill one beaker with water and the other with alcohol; and color both with the same coloring - make sure that the color intensity is exactly the same.
2. Fill the communicating tube with the colored water, so that no bubbles are found in the whole tube. (the meniscus in both legs of the tube should be level now)- see *Sketch I*
3. Put the beaker with the water away, show only the water filled communicating tube and the beaker with the colored alcohol to the audience (do not mention what the liquids are!).
4. Now pour the alcohol very carefully in one of the legs of the communicating tubes by slanting the tube and pouring the alcohol slowly - see *Sketch II*.
   What do you observe?- see *Sketch III*.

Questions:
1. Why is the meniscus in both legs of the communicating tube not level (equal in height)?
2. Why does the pouring of the liquid have to be slow?
3. What other liquids could be used instead of the water?
   Instead of the alcohol?
4. What would happen if we interchanged the alcohol and the water? How would the meniscus in both legs compare now?
5. How could we get the water levels to be even/level?
6. Will the water levels be even/level when we slant the whole apparatus?

Explanation:
This illustrates the basic principle of communicating tubes or vessels: *whenever two or more tubes, whether vertical or slanted, are connected with each other (communicating), the liquid level in all the legs will be level (same height), provided that the same liquid is being used.* This very same principle is being used to find two points that are level at the opposite corners of a house foundation. A long, clear plastic hose is almost completely filled with water, and the two ends of the hose will act as the two legs of the communicating *tubes*. In the above case, alcohol, a less dense liquid, is poured in one leg and thus the meniscus is higher in this leg, to compensate the pressure from the other leg. A denser liquid (like glycerine) would bring the meniscus in this leg lower than that in the other leg. (Compare with: COMMUNICATING CYLINDERS)

CHARACTERISTICS
OF MATTER

DENSITY
WATER PRESSURE

THE COMMUNICATING CYLINDERS

Materials: 1. One 400 or 500 ml plastic, clear graduated cylinder.
2. One 100 or 250 ml plastic, clear graduated cylinder.
3. Plastic tubing - about 30 cm /1 ft long (smallest diameter; tubing for a drip system for watering plants may be used)
4. Connecting nozzles, and a tool to puncture small holes in the cylinders.

Procedure: 1. Puncture a small hole close to the bottom, on the side of each of the graduated cylinders and insert the connectors in the holes.
2. Connect the two cylinders with the small tubing and place the cylinders on the table (see sketch on right).
3. Pour colored water in the larger cylinder and note the water levels in each of the cylinders (make sure there are no bubbles in the connecting tube by lifting one cylinder up higher.
4. Now lift the smaller cylinder (A) so high that half of the liquid runs out (water level drops back to half of what it first was). How do the movements of the cylinders themselves compare with each other?

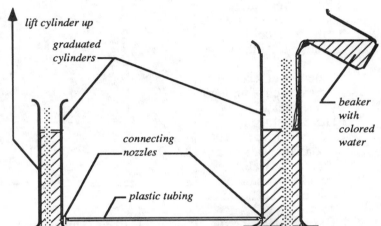

Questions:
1. Measure the distances (heights) lifted in each of the cases. Which of the cylinders had to be moved higher in order to let the same volume of water run out of the cylinder?
2. What did you notice about the water levels in each of the cylinders? If they were connected with a straight line, what kind of a line would it be? (slanted or horizontal?)
3. What did you notice about the volume of water diminishing in one cylinder and that increasing in the other? (let's say 50 ml runs out of the smaller one, by how much is the other increasing?)
4. In what ways can we make one water level stay higher than the other?
5. What would you predict if one cylinder was stoppered off before the other cylinder is raised or lowered? What would the water levels do?

Explanation:

This would be a good demonstration to assess the understanding of the *law of communicating vessels* (after having done the two former activities). The two cylinders can be considered to be the legs of a U-tube. No matter how wide the diameter of one of the cylinders, the water levels always stay at the same height (level/horizontal), because the pressure (atmospheric) on the meniscus is the same. If one leg is closed off, a lower or higher pressure results from lowering or raising the other leg and the water level in the cylinder being raised will only drop very slightly, and thus the other water level will move up slightly too. When unstoppered, moving the smaller cylinder up 10 cm will probably increase the water level in the larger cylinder with 2cm (depending on the ratio of the cylinders (practical application: *hydraulic lifts*).

In the hydraulic lift oil is being used to lift a heavy object (on the wide cylinder) with a relatively small force on the narrower cylinder. By moving the plunger in the narrow cylinder over a longer distance (with a smaller force), a much heavier object can be lifted in the wider cylinder over a much shorter distance. By doing this repeatedly (where oil comes from a large reservoir), a heavy weight like that of a car, can be lifted.

CHARACTERISTICS OF MATTER

DENSITY
WATER PRESSURE

THE PLASTIC TUBING WATER LEVEL

Materials: 1. A clear plastic hose or tubing (at least 1 m long)   2. A long carpenter's level.
3. A small beaker.   4. Food coloring.

Procedure:
1. Ask two students to hold each end of the tubing vertically so it forms a U-shape.
2. Pour some colored water (colored with the food coloring) into one of the tubing's end until there is at least 30 cm in each of the legs (see sketch).
3. Have the students hold the tubing against the board or wall, and mark the water levels (with chalk or tape on the board or wall).

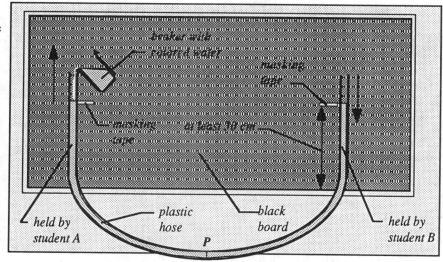

4. Have students predict what will happen to the water levels if only student A is pulling his end of the tube higher!
5. Now let student A pull his end of the tubing 10 cm higher. What happened to the water levels? Mark them on the wall.
6. Have student B bring his end of the tubing 10 cm lower. What happened to the water levels? Mark them on the wall.
7. Connect the two marks on the board or wall with a straight line and use the carpenter's level to see whether the line is level.

Questions:
1. Is the connecting line between the two marks level?
2. With how many cm did water level A go up, after student A pulled his end of the tubing 10 cm higher?
3. With how many cm did water level B go down, after student B brought his end of the tubing 10 cm lower?
4. If we connected the two marks (after A pulled it higher), what kind of line did we get? How was your prediction?
5. What would happen to the water levels if we had a bubble of air in the tubing? Try it!
6. What would the water levels do if we held the tube legs slanted? Or curved? Try it!

Explanation:
   This demonstration is illustrating the law of communicating vessels or tubes. No matter whether one leg of the tubing is pulled higher or brought lower, the water levels in both legs always stay level. Even when the legs are held slanted or curved, the water surfaces stay level. This is because the pressure on the left of point P has to stay equal to the pressure on the right of point P. See Sketch above.
   When moving the end of the tubing up on one end (say 10 cm), the water level is going up in both legs the same height, thus it is divided by two: up with 5 cm. When going down with one end of the tube, this distance is also twice as much as the water levels lowered in both legs.

CHARACTERISTICS OF MATTER

DENSITY BUOYANCY

MAKE YOUR OWN LETTER SCALE!

Materials: 1. A measuring cylinder or tall slim jar.
2. A 20-30 cm piece of broom handle, or wooden dowel or pencil with an eraser end, and 10-20 pennies.
3. A 3x5 card, thumbtack, a small metal weight (steel nut).

Procedure:
1. Tie a short thread to the weight and hook it to one end of the wooden dowel with a small nail or thumbtack.
2. Place it vertically in the water-filled cylinder to test whether it will float about half way in the water (if it floats too deeply, the weight is too big).
3. Attach the 3x5 card to the other end of the dowel with the thumbtack.
4. You are now ready to calibrate your home-made scale:
   a) Place a pencil mark at the water surface on the dowel.
   b) Put ten pennies on the card, and mark the dowel at the water surface (meniscus).
   c) Take the whole scale out of the water and divide the distance between the two marks in ten equal parts.
5. You are now ready to use the scale to weigh letters or small objects weighing less then 1 ounce or 30 grams (each line of the scale indicates .1 ounce or 3 grams: this depends on the weight of the coins used).

Questions:
1. What is the function of the small weight at the lower end of the dowel?
2. What would you say the density of wood is compared to water?
3. What other material could be used instead of a wooden dowel?
4. What variables influence the sinking distance of the dowel?
5. The capacity of the scale might be only 2 ounces or 60 grams; how can we increase its capacity?
6. How would using 2 or 3 dowels (bundled) influence the sinking distance?

Explanation:
The function of the small weight at the bottom end of the dowel is to make the dowel sink half way in the water and also to keep it floating vertically. The more dowels are used or the larger the diameter of the dowel, the greater the capacity of the scale, as the *displaced water* is increased and thus also the *buoyancy force*.

The placing of already known weights on the scale makes it possible to calibrate the wooden dowel against the sinking distance. In a less dense liquid the scale would float much lower, and in a denser liquid it would float much higher.

The use of different sizes of steel nuts will make it possible to put different weights on top of the wooden dowel (the 3 x 5" card).

CHARACTERISTICS OF MATTER — BUOYANCY MELTING POINT

THE WATER CANDLE?

Materials: 1. A drink glass or glass beaker. 2. A thick short candle, a small metal screw (or thumbtacks).

Procedure:
1. Attach the screw or thumbtacks to the bottom of the candle.
2. Fill the drink glass with water and let the candle float in it (if it sinks the screw size is too large or you used too many thumbtacks; use a smaller one or use less thumbtacks).
3. Light the candle and start asking questions.

Questions:
1. What is the function of the screw/thumbtacks in the bottom of the candle?
2. When the candle burns shorter, will the water extinguish the flame?
3. Will the candle float higher or lower in the water over time?
4. With time, will the density of the floating system change?
5. With time, will the buoyancy of the system change?
6. Will the candle actually become shorter over time?
7. When will the candle flame be extinguished?
8. What shape will the top of the candle be forming?
9. Will the candle tip over if the flame burns unevenly?
10. What would a long narrow candle do when floated in water and then lit?

Explanation:

The purpose of the metal screw or thumbtacks in the bottom of the candle is to make the candle heavier at the bottom, thus making it float in a vertical position. The top of the candle will burn leaving a concave cavity containing the molten wax. The outside of the candle which is in touch with the water stays cool and is therefore not melting. Slowly, however, the wax is burning off and the total weight/mass of the candle gets smaller.

As the mass as well as the volume of the candle become smaller, the density (mass over volume) stays the same - at least in the beginning of the burning process -, thus the part that is above water stays above water (just like 1/10th of an ice cube stays above water no matter what size of cube). But towards the end of the candle, the density becomes greater (because of the screw) and it will reach a point where the buoyancy force is not great enough to keep the candle floating.

By burning a candle this way, molten wax drippings on the outside of the candle are avoided. Almost all the wax will be burned off - a most efficient way of burning a candle!

CHARACTERISTICS OF MATTER
DENSITY OF GASES

THE FLOATING SOAP BUBBLE

Materials: 1. Soap solution (1 pt Joy detergent, 2 pts glycerine, 3 pts water).
2. A large beaker (2 litre) or clear plastic/glass container).
3. Calcium carbonate or baking soda + dilute HCl; or CCl_4; or Alka Seltzer tablets.
4. A large plastic straw, and a small cup.

Procedure:
1. Pour the soap solution in the cup and get ready to blow a medium size bubble with the straw. Put a scoop of calcium carbonate/baking soda in the large beaker and add about 50 ml dilute hydrochloric acid; or a few drops of CCl, or one or two Alka Seltzer tablets plus some water .
3. Blow a medium size soap bubble (smaller than the beaker's diameter) and let it calmly fall in the large beaker (either guide it into the beaker or move the beaker to catch the bubble).
4. Observe the soap bubble (this demonstration/activity should be performed in a draft-free room).

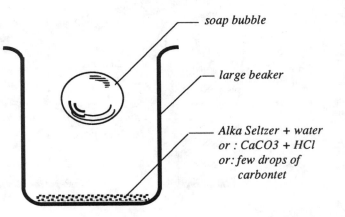

soap bubble

large beaker

Alka Seltzer + water
or : CaCO3 + HCl
or: few drops of carbontet

Questions:
1. What is being produced in the large beaker? (when using $CaCO_3$ and HCl)?
2. Why did the soap bubble float above the liquid in the large beaker and not descend all the way to the bottom?
3. How does the density of carbondioxide or carbontetrachloride gas compare with that of atmospheric air?
4. What other gas could be used instead of carbondioxide?

Explanation:
The chemicals, calcium carbonate or baking soda plus the dilute hydrochloric acid produce carbondioxide gas in the large beaker. This gas has a larger density (in other words is heavier) than air, and therefore stays in the beaker. Th e concentration of this gas is most likely higher closer to the surface of the liquid or bottom of the beaker.

The *density of a soap bubble is just slightly greater than that of air*. This is why bubbles ever so slowly descend in the air. In the large beaker the denser carbondioxide gas acts like a cushion, and this is why the bubble is not dropping all the way to the bottom.

Instead of carbondioxide gas a few drops of carbontetrachloride can be used in the large beaker. This liquid will evaporate and the gases are also heavier than air. If none of those chemicals are available, the easiest source of carbondioxide gas is one or two tablets of Alka Seltzer in a little bit of water. Baking soda and a little vinegar in the bottom of the beaker will also do to create carbondioxide gas.

CHARACTERISTICS OF MATTER

HEAT OF VAPORIZATION
VAPOR PRESSURE

THE DRINKING BIRD

Materials: 1. A glass drinking bird (supplied by Science Inquiry Enterprises or other science suppliers).
2. A drinking glass or cup (as tall as the pivot of the drinking bird)

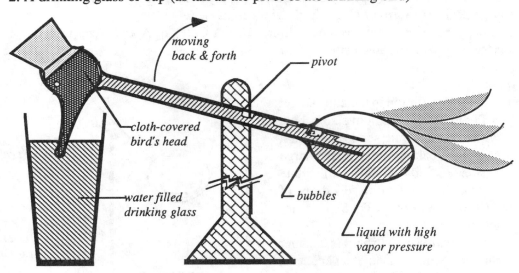

Procedure:
1. Fill the drinking glass almost full with cold water and place it in front of the bird.
2. Pick the bird up and dunk its whole head in the water for a few seconds, then place it in front of the glass, such that only its beak will touch the water.
3. Make sure that the pivot of the bird is just as high above the table as the glass rim. (If the glass is too tall, prop up the bird with a book; if the glass is too low, place the glass on a book - see sketch above).
4. Hold the bird's beak down in the water and let go to start the swinging (make sure the bird swings freely).

Questions:
1. What makes the bird keep "drinking" from the water?
2. Does the water actually flow into the bird?
3. What makes the bird to swing slower?
4. What makes the bird's head to get heavier and heavier?
5. Would the bird keep swinging ("drinking") without water?
6. Could we let the bird start swinging without water?
7. At what point will the bird stop swinging? Or will it?
8. What liquid do you think is inside the bird?
9. How would the temperature of the bird's head compare with that of its tail?
10. What function does the material around the bird's head have?

Explanation: By dunking the bird's head under water, the material around it gets soaked. When the bird swings back and forth, the water on the head evaporates, gets cooler than the rest (tail), lowering the *vapor pressure* of the liquid in the head. This lower pressure above the liquid in the head draws the liquid higher in the head, makes the head heavier and the bird "drinks" again. At the almost horizontal position a bubble of vapor passes from the tail (higher *vapor pressure*) to the head, it gets lighter and swings back, and the cycle starts over. The liquid inside the bird is most likely diethylether or any other liquid with a high *vapor pressure* at room temperature, which lowers quickly with lower temperatures.

CHEMISTRY   PHYSICAL CHANGE

WALK THROUGH A HOLE IN ORDINARY NOTEBOOK PAPER

Materials: 1. One sheet of ordinary notebook paper.   2. A pair of large, sharp scissors.

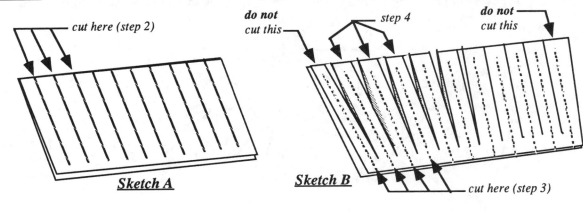

Procedure:
1. Tell the students that you are going to cut a hole in the ordinary piece of paper, and that you are going to walk right through the hole without tearing the paper!
2. Fold the paper in half, then make a series of straight cuts from the folded side, about 2 cm apart stopping about 1 cm from the edge of the opposite side (see Sketch A).
3. Turn the paper around and make cuts from the other side (see dotted lines in Sketch B), and also stop about 1 cm from the edge of the opposite side.
4. Except for the first and last strip at each end, now snip off the folded ends of the strips.
5. Then carefully open up the paper without tearing anything and walk through the hole!

Questions:
1. Do you think you could ride a bicycle through the hole?
2. We've changed the paper by cutting it; was it a physical or chemical change that the paper underwent?
3. Did we change any of the chemical properties of the paper?
4. What is the difference between a physical change and a chemical change in a sample of matter?
5. What are some other ways that you could make a physical change in this piece of paper?
6. What are some ways that you could make a chemical change in the paper?

Explanation:
   Of course, the change that we made in the paper was a physical one. No chemical properties of the paper were changed, before or after the change. When a chemical change is occurring in a sample of matter, the chemical properties in the products after the change are completely different from those before the change. To make a chemical change in the paper, it could be burned or placed in sulfuric acid. The products of the burning process of paper would be CO_2 (carbondioxide) which is a gas, plus water vapour and carbon. All three products having totally different properties compared to those of paper (the sample of matter before the change).
   Other physical changes are: the dissolving of salt or sugar in water, the stretching of a rubber band, the cutting of an apple (but as soon as the apple cuttings are exposed to the air, a chemical change takes place: browning), etc.

CHEMISTRY — OXIDATION REACTION

DRAW WITH FIRE

Materials: 1. Potassium or sodium nitrate (saltpeter).   2. A small beaker and a glass stirring rod.

Procedure:
1. Dissolve a full scoop of potassium nitrate (or sodium nitrate) in about 20 ml of water in the beaker, and stir with the glass rod until all the solid material dissolved.
2. Keep adding another scoop of KNO_3 or $NaNO_3$ and stir until the solid material stays as a solid on the bottom of the beaker.
3. Take a blank sheet of paper and draw an animal or other object on the paper with the glass rod dipped in the saturated solution. (make the lines rather thick and use the liquid rather liberally).
4. Leave the paper to dry at room temperature. Before it is completely dry, mark one spot where the drawing was started with a pencil.
5. Show the students the completely blank sheet of paper, and tell them that you will start a drawing on the paper with fire! Strike a match, blow it out, and with the glowing tip touch the marked spot on the paper; observe!

Questions:
1. Why does the pencil mark have to be made just before the drawing was completely dry?
2. What made the glowing continue further on the paper?
3. Was it actually a flame that made the drawing?
4. What kind of a reaction was it: endothermic or exothermic?
5. Is this type of a reaction a spontaneous reaction?
6. What was the glowing wood tip initially needed for?

Explanation:
Potassium or sodium nitrate is a strong oxidant. After the water of the saturated solution evaporated, there was only potassium nitrate left on the paper in a very finely divided state. The glowing tip of the match supplied the initial *activation energy* needed to release very active oxygen from the potassium nitrate. This active oxygen is further oxydizing the paper (seen in the form of a burning glow), which is exothermic (giving off heat). This energy is enough to keep further releasing the oxygen from the KNO_3. Since this is only found on the line of the drawing, the glow of the paper burn is only following the path of the drawing.

CHEMISTRY — REDUCTION REACTION

TURN A RED ROSE INTO A WHITE ONE

Materials:
1. A small red rose, and sulfur powder (S).
2. A short metal strip, and a short metal wire.
3. A glass beaker and a watchglass to cover it.

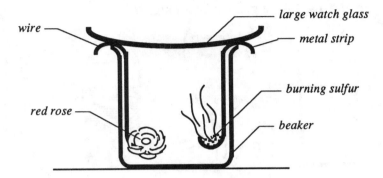

Procedure:
1. Attach the wire to the stem of the rose and let it hang inside the beaker from the rim (make a hook in the wire).
2. Bend the short metal strip in an S-shape so that it can hang from the rim of the beaker (hang this on the opposite side: see Sketch above).
3. Take this metal strip out of the beaker and place a small heap of sulfur powder on the bent part of the metal strip, hold a lit match close to the sulfur until it catches fire. Immediately lower the metal strip into the beaker and cover it with the watchglass. Observe! (Keep the beaker covered; it would be best to perform this under the fumehood or outdoors).

Questions:
1. What color did the red rose turn into?
2. What part of the rose started to change color first?
3. What type of reaction is actually taking place?
4. What gas was formed by burning the sulfur?
5. What other bleaching agents do you know?
6. What is the function of the watchglass over the beaker?

Explanation:

By burning the sulfur, sulfurdioxide gas is formed (SO_2) which is a strong reductor. The red coloring in the rose is being reduced and turns white. Sulfurdioxide is a very pungent gas and also poisonous when inhaled in large doses. This is why it is very important to dispose of the fumes under an exhaust hood or outdoors.

Other bleaching agents are chlorine compounds, like: chloride of lime, sodium chlorite, sodium hypochlorite (used in laundry bleaches) and calcium hypochlorite. In solution, these agents release chlorine gas, which reacts with color molecules and thus also remove color.

The sulfurdioxide gas plus the water in the rose petals form sulfurous acid, which works as a reducing agent. The reaction with the water is as follows: $SO_2 + H_2O \longrightarrow H_2SO_3$

| CHEMISTRY | PRECIPITATES GAS PRODUCTION |

TURN WATER INTO WINE, MILK, AND BEER!

Materials:
1. Sodium bicarbonate ($NaHCO_3$), sodium carbonate (Na_2CO_3).
2. Phenolphtalein, saturated barium chloride solution ($BaCl_2$)
3. Two drinking glasses, a wine glass, a beer mug.   4. Conc. HCl, and Bromothymol Blue.

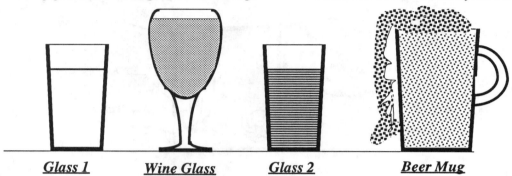

Glass 1   Wine Glass   Glass 2   Beer Mug

Procedure
1. Fill glass No. 1 three-quarters full of water and add 10 ml of saturated $NaHCO_3$ and 20% Na_2CO_3 solution (pH 9).
2. To the wine glass add a few drops of phenolphtalein.
3. To glass No. 2 add about 25 ml of saturated $BaCl_2$ solution.
4. To the beer mug add 5 ml of conc. HCl and about 3 ml of bromothymol blue. (Points 1-4 should be done without the students' knowing).
5. Now you are ready for the demonstration: Starting with glass No. 1, tell the class/audience that you have water in the glass; then pour contents into the wine glass while saying: "Now I am turning it into wine!" Now pick the wine glass up and pour contents into glass No. 2, saying: "Then I am turning it into milk!" Pick this glass with "milk" up and pour its contents into the beer mug, while saying: "Now the milk turns into beer!"

Questions:
1. What made the colourless solution turn red in the wine glass?
2. What chemicals would produce a white precipitate?
3. What reaction would produce a gas?
4. Could we turn the beer back into milk?
5. What is a one-way reaction like the above one called?
6. What reversible reactions do you know?

Explanation:
The change of the colourless liquid in glass No. 1 into the red coloured solution in the wine glass is caused by the phenolphtalein, which turns red in basic solutions. The barium ions in glass No.2 **precipitated** out in the presence of **carbonate ions** from the Na_2CO_3, and thus gave the milky white appearance. This milky mixture is actually a suspension of $BaCO_3$ solid particles in water. Pouring this suspension into the concentrated hydrochloric acid and bromothymol blue produces CO_2 gas throughout the liquid, thus forming the foam like in a beer. The bromothymol blue gave it the yellow colour at that particular pH.

This demonstration would be most effective if steps 1 through 4 were **not** shown to the students!

Invitations to Science Inquiry – Supplement – Page 58

CHEMISTRY   COMBUSTION
                                                                    PHOTOCHEMICAL REACTION

THE EXPLODING BALLOONS

Materials: 1. Three large balloons (uninflated).
2. Small gas pressure tanks of Helium, Hydrogen and Oxygen (with dispenser nozzles).
3. Thin sewing thread & masking tape.
4. A long stick with a candle taped to the end.

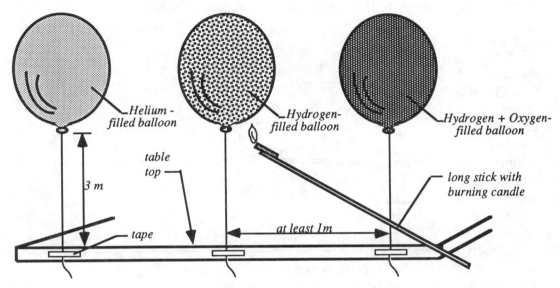

Procedure:
1. Fill one of the balloons with Helium. Let it expand to a good size and tie a knot in the balloon. Tie a thin thread to the knot and let it float to about 10 feet (3 meters) above the table top and tape the other end of the thread to it.
2. Fill the second balloon with Hydrogen and do the same, after putting a knot in the filled balloon.
3. Fill the third balloon partly with Oxygen (about 1/3 of the full size) and continue filling with Hydrogen. Tape this one also to the table top. Let the balloons float up about 3 feet (1 meter) from each other. (You are now ready for the demonstration).
4. Take the long stick with the candle, light the candle and approach each of the balloons with the flame; let students predict what will happen and ask them the questions.
(CAUTION THE STUDENTS ABOUT THE DANGERS OF COMBUSTION!)

Questions:
1. What do you think are in the balloons?
2. What can you say about the gases in the balloons?
3. What did you observe when each of the balloons bursted?
4. What can you conclude about the 3rd balloon with the very loud explosion?
5. In which of the balloons was there a physical change? In which a chemical change?

Explanation: The helium filled balloon bursts with a regular pop like a balloon filled with air. The hydrogen filled balloon will burst with a flame, since *hydrogen is combustible*. The balloon which is filled with one part of oxygen and two parts of hydrogen will burst with a very loud explosion without any visible flame. The mixture of gases is actually *photo-sensitive*, meaning that a strong bright light - like the one from the sun or a flash bulb - might trigger the reaction: $2H + O = H_2O$, creating a sudden larger volume of water vapor, resulting in the loud explosion!
    Explosions like these can be heard and detected by the human ear because of the sudden expansion of gases, which create a strong (longitudinal) wave in the air. This wave reaches the ear drum, and we hear the loud *BANG!!*

CHEMISTRY   THE AMMONIA FOUNTAIN   SOLUBILITY

Materials: 1. One round- or flat-bottomed flask (400 ml).
2. One test tube and a 1-hole stopper and bent glass tube.
3. A 2-hole stopper with a long glass tube drawn at one end in one hole and a dropper in the other hole.   4. A 500 ml beaker, a propane or alcohol burner.
5. Ammonium chloride (NH_4Cl) and calcium carbonate ($CaCO_3$).

Procedure:

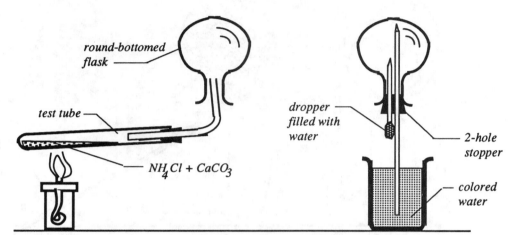

1. Mix one scoop of NH_4Cl and a scoop of $CaCO_3$ in the test tube and clamp the test tube slanted over the burner (see Sketch A).
2. Insert the 1-hole stopper with the bent tube in the test tube.
3. Fill the dropper (in the 2-hole stopper) with water.
4. Light the burner and heat the mixture of powders in the test tube while holding the flask upside down over the bent tubing (see Sketch A again).
5. As soon as ammonia vapours are detected (strong pungent smell), continue a few seconds longer with the heating, then close the flask off with the 2-hole stopper (with the long tube and dropper).
6. Immerse the end of the long tube in the beaker of water, then squirt the water from the dropper into the flask by pinching the rubber bulb (see Sketch B). Observe what is happening!

Questions:
1. What was produced by heating the two powders?
2. Why does the gas have to be collected in an upside down flask?
3. What was the function of the water in the dropper?
4. At what moment did the water rush into the flask?
5. What made the water rush up into the flask like a fountain?

Explanation:
The following reaction takes place in the test tube by heating the two powders:
$$2 NH_4Cl + CaCO_3 \longrightarrow 2 NH_3 + CaCl_2 + H_2O.$$
The ammonia gas is lighter than air and therefore it is collected in an upside down flask. As soon and since ammonia is very soluble in water, the squirt of water from the dropper was enough to dissolve all the ammonia. This made a sudden partial vacuum in the flask and the water from the beaker is therefore sucked into the flask. This demonstrates the solubility of ammonia gas in water. To make it more interesting we may put a few drops of an indicator like methyl red or bromothymol blue in the water of the beaker.

Invitations to Science Inquiry – Supplement – Page 60

CHEMISTRY — SPONTANEOUS COMBUSTION

BURN PAPER WITH ICE

Materials: 1. Sodium peroxide - Na_2O_2 2. A small chip of ice.
3. Finely chopped tissue paper, or sawdust, or starch.

Procedure:
1. Before doing anything, show students a piece of tissue paper and ask: "Would I be able to burn this piece of paper with a chip of ice?" Anticipated answer: "Impossible!"
2. Tear or cut the tissue paper into very tiny pieces and place them on a heap on an asbestos/tile plate, and build it up to a cone which is about 5 cm high in the center.
3. On top of this cone, place a half teaspoon of sodium peroxide.
4. Now show the students the small chip of ice and put it on top of the heap stand back and observe!

Questions:
1. What reaction took place? What made the paper burn?
2. What does the burning process actually need in terms of chemicals?
3. What was the function of the sodium peroxide?
4. Why was it necessary to divide the paper in such small pieces?
5. What else beside sawdust or starch can be used to replace the paper?
6. What does the ice do when left at room temperature?
7. What is the reaction between sodium peroxide and water?
8. Would this reaction be endothermic or exothermic?
9. Would regular writing paper work better or worse than tissue paper? Why?

Explanation:
The chip of ice at room temperature will melt and turn into water. The reaction between water and sodium peroxide is as follows: $Na_2O_2 + H_2O \longrightarrow 2NaOH + O_n +$ energy (heat)

The oxygen released from the above reaction is in **status nacendi**, this means that it is in atomic form, and thus very reactive. The very reactive oxygen immediately attacks the combustible paper snippers and sets it into flame. By chopping the paper into fine small snippers, we actually decrease the ***kindling temperature.***

In place of the paper we can use fine sugar, lycopodium powder, fine coal dust, or any other easily combustible material. The finer the combustible material, the lower the kindling temperature. The decomposition reaction above is very ***exothermic***, and the released heat is enough to decompose more of the sodium peroxide, which in turn releases more active oxygen.

The ice (or water) in this case functioned as an ***initiator,*** as it was only needed to release the oxygen from the sodium peroxide in the beginning. It was <u>not</u> acting as a catalyst, since it participated in the reaction.

CHEMISTRY   SPONTANEOUS COMBUSTION
                                             GAS PRODUCTION

THE POTASSIUM CHLORATE BOMBS

Materials: 1. Potassium chlorate, white phosphorus (kept under water).
2. Carbon disulfide, Kleenex tissue.   3. Asbestos plate (or tile).
4. Small beaker and medicine dropper.

Procedure:
1. Make a solution of white phosphorus in carbon disulfide, by cutting a small chunk off the white phosphorus stick (hold with tongs) and placing it in some carbon disulfide in the small beaker (a half cm cube of phosphorus in about 10 ml of carbon disulfide).

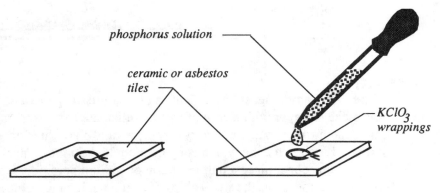

2. Wrap a very small amount of potassium chlorate in tissue paper and tie it off with a thread, such that it makes pea-size wrappings (cut the excess tissue paper off).
3. Place these little wrappings on the tile or asbestos plate, one on each tile and place the tiles about 10 cm apart from each other.
4. Now, with the dropper place two or three drops of the phosphorus solution on each of the wrappings, and stand back!

Questions:
1. What triggered the sudden explosions?
2. What does the phosphorus do when dissolving in the carbon disulfide?
3. What kind of liquid is carbon disulfide?
4. What do we call a liquid which evaporates quickly?
5. Why is white phosphorus kept under water when stored?
6. What happens to the Phosphorous-solution after dropping it on the wrappings?
7. What gas is released by the potassium chlorate?
8. What is it that makes us hear an explosion?

Explanation:
   White phosphorus is very reactive in the presence of oxygen. This is the reason why it is kept under water when stored. Dissolving it in carbon disulfide, which is a very **volatile** and **combustible** liquid, breaks it up in very small, atomic particles. After dropping the solution on the wrappings of potassium chlorate, the volatile (easily evaporating) carbon disulfide evaporates and atomic phosphorus is left on the tissue paper. The phosphorus reacts with the oxygen in the air, releasing heat, and heat causes the breakdown of the potassium chlorate into potassium chloride and atomic oxygen: $KClO_3 \longrightarrow KCl + 3O_n$. Thus a sudden formation of gases is the result, causing a sudden large wave in the air, which our eardrum senses in the form of a BANG!

CHEMISTRY — SPONTANEOUS COMBUSTION

BURN A PIECE OF METAL IN WATER

Materials: 1. Sodium metal (kept under liquid paraffin or oil).
2. A petri dish, and phenolphtalein.
3. An overhead projector (for large audiences).

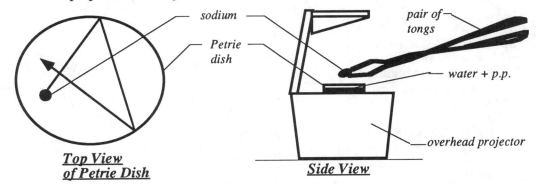

Top View of Petrie Dish   *Side View*

Procedure:
1. Fill the petri dish about half way with water and place it on the overhead projector. Add a few drops of phenolphtalein and stir.
2. Cut a <u>small</u> chunk of sodium metal (use tongs to take the sodium out of the liquid, dry the sodium off on a paper towel and cut a piece off with a knife) and place the rest of the sodium back in the liquid.
3. Pick the piece of sodium up with the tongs and drop it in the water in the petri dish on the overhead projector (with an outstretched arm). Let students observe the screen as well as the dish from the side!

Questions:
1. What did the sodium do after it was dropped in the water?
2. What did you notice about the colour of the water after the dropping?
3. In what environment does phenolphtalein turn reddish pink?
4. What do you expect the pH of the water to be after the dropping?
5. What are the products of the reaction of sodium and water?
6. What was the cause of the flame? (Not every time is a flame created).
7. Why did the sodium piece make such erratic motions over the water?
8. Why would it be dangerous to use too large a chunk of sodium?

Explanation:

Sodium metal is so reactive that it has to be kept under liquid paraffin. If it is kept in the open air, it will react with the water vapour in the air. Dropping a piece of sodium in water will give the following reaction: $2Na + 2H_2O \longrightarrow 2NaOH + H_2(gas)$. This is an *exothermic* reaction, and since hydrogen gas is flammable, the heat produced is high enough to ignite the hydrogen. The sodium hydroxide produced gives the water a basic reaction. The pH gets to be higher than 7 and the phenolphtalein makes the solution to turn red.

Since the hydrogen gas is produced more on the contact point with water on one side than on the opposite side of the piece of sodium, a movement of the piece is caused in the water. This movement is eratic and random because the sodium piece is irregular.

Too large a chunk of sodium in water might produce sudden large amounts of hydrogen gas, which may explode when brought in contact with the water. Therefore this is very dangerous!

CHEMISTRY

COMBUSTION
EXOTHERMIC REACTION

THE DELAYED EXPLOSION

Materials: 1. Tin can with a tight tin lid.   2. Bunsen burner (or propane burner) & tripod stand.

Procedure:
1. Make small holes in center top and bottom of tin can.
2. Fill the can from the bottom hole with gas from the Bunsen or propane burner (this takes about 30 sec).
3. Light the top flame by holding a lit match over the opening of the top hole.
4. Once the flame is on, take the burner away and shut it.
5. Wait and stand back .... EXPLOSION .... lid flies off!

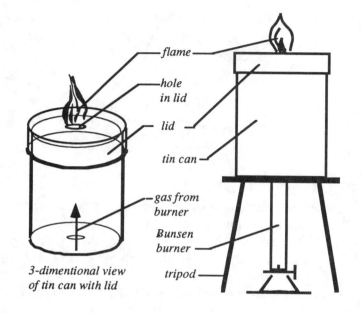

3-dimentional view of tin can with lid

Questions:
1. What makes the lid fly off?
2. Why don't we get an explosion at the moment that we were lighting the top flame?
3. What would happen if we didn't wait for the can to fill up with gas, and light it?
4. What would happen if we carried out the whole event with the can upside down?
5. What other gases can be used to do this demonstration?

Explanation:

Natural gas is a mixture of methane (CH_4) and hydrogen (H_2) gas. When the can is totally filled with the gas, it can be lit at the top opening. This flame will be sustained as long as the gas is being supplied. As soon as the gas supply is shut off and the burner is taken away, air comes into the can from the bottom. The air mixes with the natural gas and as soon as there is enough oxygen present, an *explosive mixture* is formed and the flame above the can sets it off: EXPLOSION .... and the can's lid flies off!

This can happen when a propane or natural gas leak occurs in a home or factory. An explosive mixture of the gas and air is formed, which is set off by a small spark or flame. The optimum mixture of gases to become an explosive one is: hydrogen : oxygen = 1 : 2. This means that the proportion in terms of natural gas to air is as 1 : 10, because of the 20% oxygen content in air. Only 10% of gas contaminated air is therefore already very explosive!

Since an explosive mixture of gases can so easily be obtained, the filling of the can with the combustible gas must therefore be complete before a burning match is held above the opening in the lid. If we do not wait until the can is full of gas, an immediate explosion might occur. How can we know that the tin can is already full of gas? While the can is being filled with the gas, bring some of the escaping gas towards your nose by waving with your hand. When the smell of gas can be detected, it is an indication that the can is already full.

CAUTION: NEVER BRING YOUR NOSE OVER THE OPENING TO SMELL THE GAS!

Invitations to Science Inquiry – Supplement – Page 64

CHEMISTRY   COMBUSTION
                                                               KINDLING POINT

THE STRAW FLAME THROWER

Materials: 1. A plastic spoon straw (or regular drinking straw)
  2. A candle & matches.   3. Lycopodium powder.

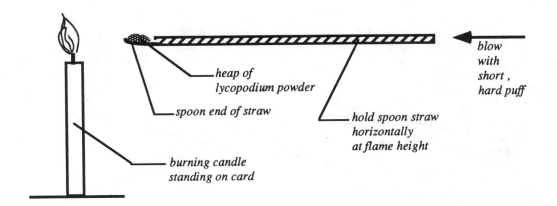

Procedure:
1. Light the candle and attach it to the table top or a 3x5 card (with a few drops of molten wax).
2. Take the spoon straw or make one from a regular straw: cut about 1/2 inch into the straw with a pair of scissors, then cut perpendicular to the straw about half way and cut a spoon shape out of the cut piece (see above sketch).
3. Dip the spoon in the lycopodium powder such that the spoon has a heap of powder on it.
4. Hold the straw horizontally and direct it towards the flame of the candle: hold the spoon about 2 inches from the flame and put the other side of the straw in your mouth and blow a short hard puff into it.
5. A variation to this would be: to place some lycopodium powder on your right palm and a lit match in your left hand; throw the powder into the flame!

(CAUTION THE STUDENTS ABOUT THE DANGERS OF COMBUSTION!)

Questions:
1. What did you observe?
2. Where did all the powder go?
3. Will this also happen with regular flower or starch?
4. What can you conclude about the combustibility of lycopodium powder? About its kindling point?
5. Will the powder catch flame when the match flame is brought near a heap of the powder?
6. How can we lower or increase the kindling point of a certain material?

Explanation:
   The physical condition of lycopodium powder is such that it is very finely divided. It consists of spores of club moss. Because it is so fine and very dry, its kindling point is very low; and since there is lots of air (or oxygen) around the particles, a sudden complete combustion takes place. When we try this with regular flower or starch, we may succeed in getting combustion, but since the particles are usually larger and somewhat more moist, only a partial combustion is achieved. Grinding matter very finely will decrease the *kindling point* - making it more combustible - and conglomerating it will increase this point - making it less combustible. Sparks in flower bins or coal mines may cause explosions and water sprayers will usually diminish this danger.

CHEMISTRY   FLAME AND BURNING

THE HUMAN FLAME THROWER

Materials: 1. A sheet of paper (towel or writing paper).
2. A candle and match.
3. A container of water.

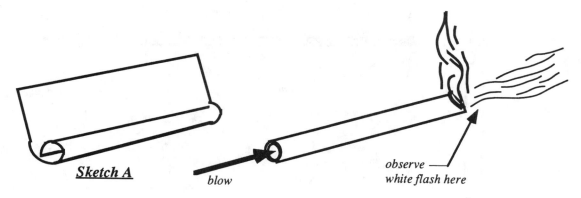

Procedure:
1. Light the candle and attach it to the table top.
2. Roll up the sheet of paper rather tightly in a cylinder of about 3 cm diameter (see Sketch A).
3. Light one end of the paper roll with the burning candle and hold the roll horizontally with the burning end slightly upwards.
4. Without touching your lips to the other end of the paper roll, blow a short puff through the roll (the blowing can be repeated).
5. Observe a brighter flash of light and listen to the sound it makes!
6. Dump the burning paper in the container of water.

Questions:
1. What makes the flame move outwards?
2. What causes the brighter flash of light?
3. What kind of sound did the flame make during the blowing?
4. Why must we hold the roll with the burning end slanted upwards?
5. What would happen if the burning end was held lower than the other end?
6. Why is it better to hold your lips a little away from the roll of paper when blowing through it?
7. What are the products of the burning process?
8. What are the elements that paper molecules are made up of?

Explanation:
   After burning one end of the paper roll, it is better to hold the burning end slightly higher than the other end, so that the smoke formed by the burning is flowing up and not through the roll to your mouth side. This demonstration works best if the roll was held perfectly horizontal. This way the smoke will be kept inside the roll and at the time that we blow, the smoke comes out the other end. This smoke mixed with the air burst into flame and gives a bright flash of light. At this moment we hear a ...pllaap...sound, indicating the sudden formation of gases.
   Paper is composed of plant fibers and these consist of cellulose molecules: $C_{12}H_{22}O_{11}$ and when burned gives off CO_2 (carbondioxide) gas, and water vapour, and leaves behind the black material, which is carbon (C).

CHEMISTRY   FLAME AND BURNING

THE MAGIC CANDLE

Materials: 1. A medium size candle.   2. A book of matches.

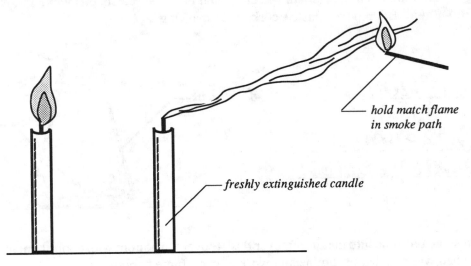

hold match flame in smoke path

freshly extinguished candle

Procedure:
1. Light the candle and let it burn until the flame becomes a good size.
2. Strike another match. While holding the burning match in your hand, blow out the candle flame with a short puff.
3. Immediately after the flame is out, hold the flame of the burning match in the smoke trail coming from the candlewick (about 3-4 cm from the wick).
4. Observe the travelling flame! (If smoke trail is not thick enough, repeat points 1, 2, and 3 again. A draft-free room is needed to carry this out).

Questions:
1. What makes the flame travel to the candlewick?
2. Why does this demo have to be carried out in a draft-free room?
3. What is actually burning of the candle?
4. Would wax without a wick sustain a flame?
5. What else beside a wick can we use to make wax sustain a flame?
6. What creates the smoke after a candle is blown out?

Explanation:
   What is actually burning of a candle is the vapor which comes from the heated candlewax. When a candle is being lit, it is the wick that burns first. The heat of the flame melts the wax, the wick absorbs the molten wax and the hot wax vapors burn around the wick. When blowing out the candle flame, the hot wick keeps heating the molten wax. This is the reason why a smoke trail is formed. This smoke actually consists of very small **dispersed particles of wax** in the air, making it highly combustible. Holding a flame within this smoke trail sets it aflame and the flame "travels" to the wick.
   The "Magic Candles" that can be bought in confectionary stores are candles which cannot be blown out. After blowing the flames out of these candles, they light themselves again. This is because the wax is treated with red phosphorus. The glowing wick will make the little particles of phosphorus spark up and make the candle burst back into flame.

CHEMISTRY — REACTION ENERGY

THE GLOWING ALUMINUM

Materials: 1. Aluminum chunks (small pea size), and bromine (poisonous vapours).
2. A 300-400 ml Erlenmeyer flask + cork, wide masking tape.

CAUTION:

HAZARDOUS HALOGEN FUMES!

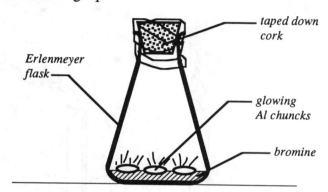

Procedure:
1. Pour some bromine into the Erlenmeyer flask (cover the bottom with about 1 cm layer. Use gloves when working with bromine and work under a fume hood or outdoors!)
2. Open the cork and drop two or three aluminum chunks in the bromine. Stopper the flask immediately and tape the cork down securely.
3. Darken the room and observe the glowing aluminum!

Questions:
1. What made the aluminum glow?
2. What particular reaction was taking place?
3. Did the reaction give off or take in energy?
4. Why did the stopper have to be taped down to the flask?
5. Would other halogens also react the same way with aluminum?
6. Which of the halogens would react faster or slower with aluminum?
7. How can we promote contact between two substances that are both solids?

Explanation:
 The reaction between aluminum and bromine is exothermic. This means that it is giving off energy in the form of heat. This heat makes the aluminum glow in the dark. The reaction between aluminum and bromine is as follows: $2\ Al\ (s) + 3\ Br_2\ (l) \longrightarrow 2\ AlBr_3\ (s) + energy$
 Other halogens are chlorine and iodine, and both will react with aluminum. Chlorine is a gas and will react much faster with aluminum than bromine. Iodine reacts much less readily with aluminum. In order to let them react with each other, both solids have to be divided in very small particles, in other words they have to be very finely ground. When mixed with each other, they still do not react, but a tiny drop of water will start the reaction (see THE FIERY WATER).
 When working with halogens we have to be very careful and always be very cautious in handling them. Always work under a fumehood or outdoors. The fumes of all three halogens are poisonous and are **very hazardous** for our pulmonary system, so never inhale even the slightest amounts of halogen fumes!

CHEMISTRY   REACTION ENERGY

TOUCH A CRACKER!

Materials: 1. Concentrated ammonia - $NH_{3(aq)}$ Conc.
2. Iodine (solid crystals).   3. A meter stick, filter paper, a balloon, a pin.

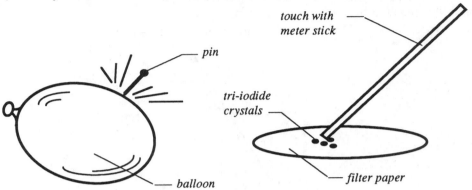

Procedure:
1. Dissolve some iodine in concentrated ammonia; make sure that all solid material disappears (use a glass stirrer and a small beaker).
2. Let the solution stand in the air for a while until some solid brown material is formed on the bottom of the beaker.
3. While this solid material is still wet, transfer it with the stirrer to a filter paper and let it dry in the air (this will take about an hour to completely dry; the drier the better).
4. Place these dry crystals about a meter away from you, let students stand back, hold the meter stick at the end and tap the other end of it on the crystals.
(CAUTION: TAP ONLY A FEW CRYSTALS AT A TIME: VERY EXPLOSIVE!)

Questions:
1. What compound was formed by dissolving the iodine in ammonia?
2. What is it that makes a chemical so explosive?
3. What are other chenicals or chemical mixtures that are explosive?
4. What makes us hear the loud noise in an explosion?
5. What is always suddenly formed in an explosion?
6. Does an explosion always have to be accompanied with a fire?
7. Does oxygen always have to be present for an explosion to occur?
8. What can we compare the trapped energy in the nitrogen tri-iodide with?
9. How does a bursting air balloon compare with this explosion?

Explanation:
The wet nitrogen tri-iodide is still safe to handle, but as soon as all liquid is evaporated (usually excess ammonia), the dry solid is very unstable and explodes the moment it is touched or walked on. It is harmless when very small quantities at a time are used.

Nitrogen tri-iodide can be compared to a set mousetrap, and the touching of it to the releasing of the trap. It is an unstable compound that suddenly forms nitrogen gas when it is mechanically upset. This sudden formation of gas is the cause for us to hear the explosion. Explosions are always accompanied with the sudden formation of gases, whether it is in the formation of water vapor from its elements, or the igniting of explosive mixtures (See Events 5.7; 5.26; and 5.28 of ISI 2nd Ed).

CHEMISTRY — ENDOTHERMIC REACTION

THE STICKY BOARD

Materials:
1. A small beaker (250 ml), glass or plastic stirrer.
2. A small thin wooden board (10 cm x 10 cm)
3. Barium hydroxide - $Ba(OH)_2$ and ammonium thiocyanide - NH_4SCN

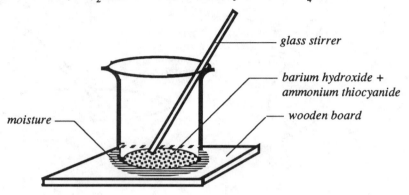

Procedure:
1. Wet one side of the wooden board thoroughly with water.
2. Place the beaker on the wet side of the board.
3. Add one teaspoon of each of the chemicals to the beaker and stir while holding the beaker down tight to the board.
4. After the two chemicals are thoroughly mixed, lift the beaker slowly up (formation of frost on the outside of the beaker is an indication that it is ready to be lifted).

Questions:
1. Why did the wooden board have to be wetted?
2. What kind of reaction occurred between the two chemicals?
3. Why did the wooden board stick to the beaker?
4. How else can we show that the reaction took away heat?
5. What other types of reactions do you know?

Explanation:

Most reactions are exothermic, which means that the reaction gives off heat (See Events 5.23 - 5.29 ISI-2nd Ed), but this particular one is *endothermic*. This means that the reaction takes away heat, in other words the reaction requires calories in order for it to occur and if this heat is not supplied, it withdraws the heat from its environment. This cools down the environment and thus freezes the moisture that is between the beaker and the wooden board. This is the reason why the wooden board sticks to the beaker after the two chemicals are mixed in it. The reaction between the two chemicals is as follows:

$$Ba(OH)_2 + 2NH_4SCN \cdot (H_2O)_4 \longrightarrow Ba(SCN)_2 + 2NH_4OH + 8H_2O - energy$$

We can see that the reaction gives off water also besides taking away energy, this is why the two dry chemicals become quite watery after mixing. It is therefore not necessary to add water to the reactants in the beginning of the mixing.

CHEMSTRY — REVERSIBLE REACTION

SHAKING THE BLUES

Materials:
1. An Erlenmeyer flask (250/400 ml) and stopper.
2. Potassium hydroxide (5g), Glucose or dextrose (3g).
3. Methylene blue.

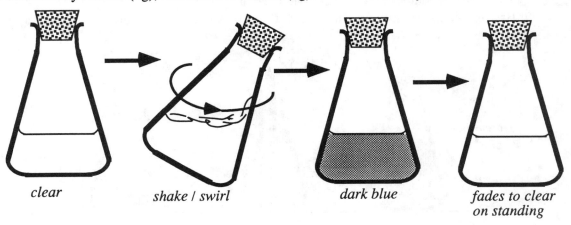

clear → *shake / swirl* → *dark blue* → *fades to clear on standing*

Procedure:
1. Dissolve the above chemicals in the Erlenmeyer flask with 250 ml of water.
2. Show to the students that it is a clear colorless liquid.
3. Stopper the flask and shake the solution vigorously and show the change of color.
4. Place the flask on the table and leave it standing.
5. Draw students' attention to the color of the solution.
   (The shaking may be repeated to show the color change for several hours).

Questions:
1. What type of reaction is this? One that can be repeated seemingly almost endlessly?
2. When will the turning into the blue color stop? Or will it ever stop?
3. What makes the clear, colorless liquid turn blue?
4. What can we do to make the liquid turn blue again? (after it has stopped turning color?

Explanation:
Methylene blue is reduced to a colorless compound by an alkaline solution of a reducing sugar: in this case the glucose or dextrose. When we shake the flask, the colorless solution is reoxydized by the oxygen above the liquid into the blue dye: methylene blue. When it is left standing, it is being reduced again, thus the gradual fading of the blue color. Characteristic of a *reversible reaction!*

Slowly, however, the oxygen in the flask above the liquid is seeping through the cork and is also reacting with the reducing sugar. This is why it eventually (after a few hours) stops changing into the blue color. To make the solution turn color again, the oxygen has to be replenished. This can be done by leaving the flask open to the air for a few moments.

After a few days standing, the solution will turn yellow and to brown. A freshly prepared solution should be used for this demonstration.

Most chemical changes are **irreversible,** this means that it can only proceed in one direction, like the burning of a candle. All physical changes, on the other hand, are almost always **reversible,** meaning that it can go back to the original materials, like the dissolving of sugar in water.

CHEMISTRY   ORGANIC CHEMISTRY
                                                                   UNSATURATED BONDS

THE COLOR ABSORBING BACON

Materials: 1. Three or four strips of crisply fried bacon.
2. Bromine in an Erlenmeyer flask (250/400 ml), stoppered with cork.

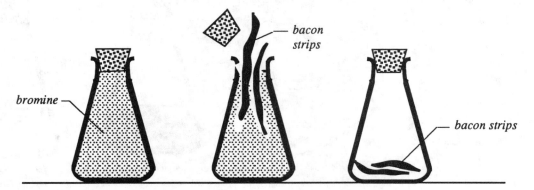

Procedure:
1. Place a few drops of bromine in an Erlenmeyer flask and quickly stopper it with a cork (make sure that the cork fits the flask prior to adding the bromine to the flask!).
2. Show students the brown color of the bromine vapours, and tell them that in compound form, that is if they react with other chemicals, that there is no characteristic of bromine vapour left.
3. Open the stopper of the flask and quickly drop the bacon strips into the flask, close again and shake.
4. Hold a white sheet of paper behind the flask and let students observe the color change!

Questions:
1. What color did the brown vapour turn into?
2. What made the brown color disappear?
3. What do double bonds do when a halogen is added to it?
4. What other halogens would behave the same way as the bromine?
5. How do we know whether all the bromine has reacted?
6. How do we know whether all the unsaturated bonds have reacted?
7. What are unsaturated bonds in organic compounds?

Explanation:
Crisply fried bacon has a high content of unsaturated fats. These are organic compounds containing double or triple bonds between the carbon atoms. When halogens are added to these compounds, the halogens are active enough to open these bonds and attach themselves to the organic molecule. See reaction below:

$$Br_2 + -C=C- \longrightarrow \begin{array}{c} H\ H \\ |\ \ | \\ -C-C- \\ |\ \ | \\ Br\ Br \end{array} \quad \text{or} \quad 2Cl_2 + -C\equiv C- \longrightarrow \begin{array}{c} Cl\ Cl \\ |\ \ | \\ -C-C- \\ |\ \ | \\ Cl\ Cl \end{array}$$

brown gas   colourless   green gass   colorless

These reactions are called **addition reactions.** Iodine vapours would most likely also react with the bacon strips, except that it would take a little longer, as iodine is the least reactive of the three halogens.

Invitations to Science Inquiry – *Supplement* – *Page 72*

CHEMISTRY

ORGANIC CHEMISTRY
HIGH POLYMERS

MAKE A NYLON THREAD OUT OF TWO LIQUIDS

Materials: 1. Hexamethylene diamine (1,6 hexanediamine) $NH_2(CH_2)_6NH_2$
2. Sebacyl chloride (1,10 decanedioxyl chloride) $COCl(CH_2)_8COCl$.
3. Carbontetrachloride (tetrachloroethane).
4. Sodium carbonate, 2 beakers, tweezer, stirrer.

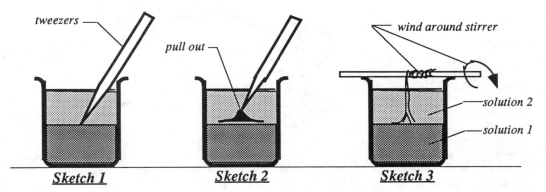

Sketch 1   Sketch 2   Sketch 3

Procedure:
1. Make in a 100 ml beaker a solution of 2 ml sebacyl chloride in 50 ml of carbontetrachloride.
2. Make in another beaker a solution of 2 g of 1,6 hexanediamine and 4.0 g of sodium carbonate in 50 ml of water.
3. Pour this second solution very carefully on top of the first solution by slanting the first beaker and slowly pouring the second solution over the first one.
4. With the pair of tweezers, pick up the center interface between the two liquids and pull it slowly out of the liquid (see Sketch I and II).
5. Then wind this thread around a glass stirrer and keep turning the glass rod to continue pulling the nylon thread out.

Questions:
1. What compound is formed at the interface of the two liquids?
2. What would happen if the top liquid was just poured carelessly into the first solution?
3. What would happen if the two liquids were stirred together?
4. Is the formed compound identical to the industrial Nylon 66?
5. How long do you think the thread can be pulled?
6. At what point would the thread stop forming?

Explanation:
The liquid containing sebacyl (or adipyl) chloride and the other liquid containing hexamethylene diamine react at the interface to form Nylon-610.

$$NH_2(CH_2)_6NH_2 + COCl(CH_2)_8COCl \xrightarrow{-HCl} -N(H)-(CH_2)_6-N(H)-C-(CH_2)_8-C-N(H)-(CH_2)_6-N(H)-C-(CH_2)_8-$$

hexamethylene   **sebacyl chloride**   **Nylon-610**

The above reaction is a *polymerization* reaction. Large and extremely long molecules are formed. This same polymerization occurs in the production of high polymers like, polyethylene, polystyrene, polyurethane, and other plastics.

CHEMISTRY   TECHNOLOGICAL
                                                     APPLICATION

THE STICKY MATCHES

Materials: 1. A box of wooden safety matches.

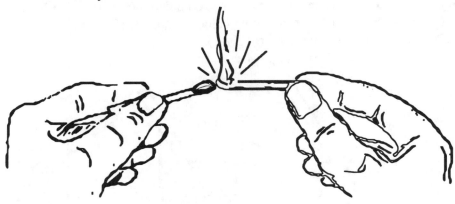

Procedure:
1. Take two match sticks out of the box and hold one in each of your hands.
2. Strike the one in your right hand against the box (if you are right handed) to light the match.
3. Immediately hold the lit match head against the match head in your left hand and wait till the left match gets lit.
4. As soon as the hissing sound of the ignition stops, blow out the flame. Make sure that you are holding the two matches with a steady hand against each other (see Sketch above).
5. Slowly let go of the match in your right hand (without moving/jarring the match - if you don't succeed the first time, try again!)

Questions:
1. What made the two matches stick together?
2. What would happen if the first match was put against the second <u>not</u> immediately after it was lit?
3. How are matches produoed?
4. What are matches made of?
5. What is the match head made of?
6. What makes a match head burst into flame when rubbed against the box?
7. What chemical is on the striking surface of the box?

Explanation:
   The red tips of the matches burn and fuse together, making the matches stick. The burning of the red tips form chemicals that are porous and rough in texture. These porous and rough tips "grab" each other and thus making the matches stick together.
   The match stick itself is made of softwood veneer. The wood splints are soaked into a bath of sodium silicate, ammonium phosphate or sodium phosphate and then dried. This impregnation prevents the afterglow. The wood splint is then dipped in a paraffin bath, which sustains burning. The match head oonsists of an oxygen carrier (potassium chlorate, chromate or lead oxide), sulfur and abrasives like powdered glass, and binding agents (dextrin or gums).
   The side of the box contains red phosphorus and powdered glass. Safety matches can only be lit by striking against the side of the box. The friction produces heat, which releases the oxygen from the potassium chlorate and this reacts with the sulfur to produce the flame.

CHEMISTRY — COMPLETE COMBUSTION

MAGIC FLASH PAPER

Materials:
1. White tissue paper
2. Concentrated sulfuric acid
3. Concentrated nitric acid
4. Two dinner plates & 2 glass rods

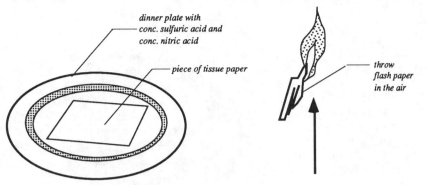

Procedure:
1. Mix four parts of concentrated sulfuric acid with five parts of concentrated nitric acid on the dinner plate (mix small amounts at a time) and stir well with the glass rod.
2. Let the mixture cool off till it reaches room temperature.
3. Immerse the tissue paper in the acid mixture for about 10 minutes.
4. Take the tissue paper out with the glass rods and rinse it under running water.
5. Place it on the second dinner plate and let it air dry.
6. Once the paper is dry, take small pieces at a time (2 x 2"), burn a corner and throw in the air!

TO PRODUCE COLORS:
1. Prepare separate saturated solutions of the following (use as much chemical as can dissolve in the water):
    For RED flash paper...........................Strontium nitrate
    PALE GREEN flash paper.............Barium nitrate
    DEEP GREEN flash paper.............Copper chloride
    VIOLET flash paper......................Potassium nitrate
2. Soak small pieces of the prepared flash paper (before or after it is completely dry) in the different salt solutions, and let completely dry.
3. Crunch up the small pieces, burn at a corner and throw in the air!

Questions:
1. What was the function of the concentrated acids?
2. What would happen if we only used concentrated sulfuric acid?
3. What other chemicals are usually added in the production of paper?
4. What is usually left over after burning a piece of paper?
5. In order for the paper burning process not to leave anything solid, what do the end products of the chemical reaction have to be?
6. What made the flame to turn red? Pale or deep green? Yellow? Violet?

Explanation:
In the production of paper from wood pulp, **fillers** are added, like: Calcium carbonate to make it heavier, and calcium hypochlorite to make it whiter, and many other chemicals. When we burn regular paper there is always ash (solid material) that is left over, mostly caused by the fillers.

The purpose of the concentrated acids is to dissolve and remove the fillers in the paper. When we remove all the fillers from the paper, only cellulose is left over, and since cellulose has the general formula of $C_5H_{10}O_5$, it can easily be burned and changed into CO_2 and H_2O. A **complete combustion** is therefore taking place when we burn the flash paper, meaning that there are no ashes left over.

The different colors are caused by the different elements present during the combustion: Strontium gives a red color, Barium a pale green color, and Copper a deep green color, Potassium a violet color, and Sodium which is usually already present in the paper, gives a yellow color in the flame.

Invitations to Science Inquiry – Supplement

ENERGY   NUCLEAR SOURCES

THE SUN-BAKED POTATO

Materials: 1. A round bread basket or half a coconut shell. 2. A small raw potato.
3. Aluminum foil (enough to line the basket). 4. A toothpick.

Procedure:

1. Take a piece of aluminum foil large enough to line the basket from the inside, and cover the inside of the basket with the foil - make sure the shiny side is showing.

2. Place a small raw potato in the center of the basket, and place the whole basket in the sun (do this preferably at a time when the sun shines at its hottest - between 11 am and 3 pm).

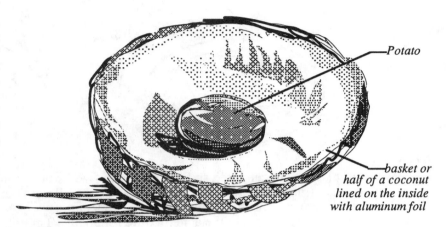

3. Leave this to bake for about 30 - 45 minutes and check the potato with the toothpick for softness (if the potato is done, it should be soft enought to pierce with the toothpick).

4. Pick a day that is wind still to do this. If there is some wind, place a cardboard shield around the basket. Make sure that no shadow falls on the basket!

Questions:
1. What is the function of the aluminum foil?
2. Does the shape of the basket have to be round? What other shapes can be used?
3. How can we make optimum use of the reflected sun rays?
4. At what spot do the sun rays come together? What is this spot called?
5. Would a larger or smaller basket be more effective in heating the potato?
6. Why do we carry out this event close to noon time? Why not early morning or late afternoon?
7. What process is actually happening on the sun to give off that much energy?
8. What forms of energy were the source and the end product?
9. What other ways are there to concentrate the energy from the sun?

Explanation: The original source of energy used is the sun's energy, which is nuclear, where four Hydrogen atoms are *fused* into one Helium atom. During this *fusion* process energy is radiated out in the form of sun rays. Only a small fraction of these sun rays falls on the earth, and again a very, very small fraction is falling on the basket. These sun rays are concentrated into a much smaller area by the shape of the basket and the reflecting aluminum foil, exactly where the potato is placed.

The final form of energy is heat, which is cooking the potato. The larger the basket, the more sunrays it is catching, and thus the more effective it is in heating the potato. A basket in the shape of a parabolic disc would be the ideal shape, but the potato has then to be raised into the *focal point* of the disc. This can be achieved by placing the potato on a wire tripod. (see Sketch on right).

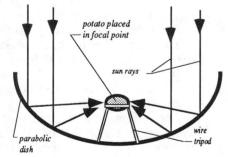

The sunshine is hottest around noon time, because at that time the sun rays fall on the earth almost perpendicular to the surface of the earth. Early in the morning or late afternoon the sun is closer to the horizon, and the sun rays strike the earth surface in an angle. This makes the number of rays per square area much less, and thus the intensity of the sun is not as severe.

ENERGY SOLAR SOURCES

THE TEST TUBE GREENHOUSE

Materials: 1. Two thermometers (0-100°C). 2. A large, wide test tube or long jar with a narrow mouth.
3. One-hole stopper to fit the thermometer in the test tube.
4. A white or infra red spot light.

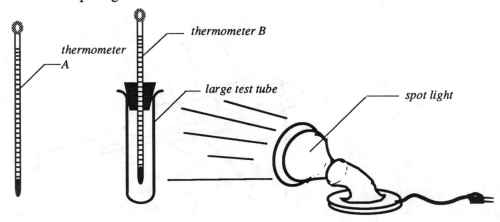

Procedure:
1. Insert one thermometer in the one-hole stopper and place this over the large test tube or jar (if a loose thermometer is not available, a paper covered jar may be used instead of the large test tube and the stopper).
2. Place both thermometers in the sunlight or shine the spot light on them from about 50 cm away (see Sketch above).
3. Have the students read and record the temperatures of both thermometers every minute for about 15 minutes.

Questions:
1. Which of the thermometers shows a higher temperature?
2. What is the light heating around thermometer A, and what is it heating around thermometer B? (see Sketch above)
3. What do these results tell us when applied to a sun room or greenhouse?
4. Which direction should these sun rooms face when your home is located in the Northern hemisphere? In the Southern hemisphere?
5. Would the temperature on thermometer B be higher or lower if the tube or jar were made of opaque glass? If the tube or jar were covered with white paper? If it were covered with black paper?

Explanation:
Like the atmosphere of the earth, glass around the thermometer can trap heat energy. The light rays heat up the air in the tube, which cannot move around, contrary to the air around thermometer A. Similarly, sunlight energy is trapped in the atmosphere, as the sun's energy is absorbed by the earth. Part of this energy is reradiated into the atmosphere but cannot pass out of the atmosphere.

A *greenhouse* works on the same principle. If the glass walls of a greenhouse were covered with black material, it would make the room much hotter, but since plants need light, glass is needed. Other examples of energy traps are: closed cars left in the sun, especially the dark colored ones; attics of homes with black roofs, etc.

ENERGY   SOLAR SOURCES

START A FIRE WITH A MAGNIFYING GLASS

Materials: 1. A large magnifying glass (or strong reading glasses).
2. A bright sunny day, over-exposed film or dark glasses.
3. A piece of tissue paper (or any easily combustible material).

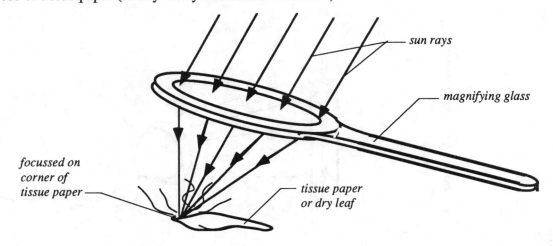

Procedure:
1. Take the class outdoors on a bright sunny day, and bring the large magnifying glass or a pair of strong reading glasses with you. You also should have a piece of tissue paper or any thin paper in your pocket.
2. Place the tissue paper on a spot where everybody can see it. Let the students use the darkened film or dark glasses to observe the paper, while you hold the magnifying glass about 20-30 cm above the paper. This distance has to be varied by moving the lens up or down, such that the sun rays are focussed on the paper. (The small bright circle has to become a small point if it is well focussed).
3. Hold the magnifying glass at that position (focussed position) for awhile (about 15-20 sec) until suddenly the paper starts to smolder and burn.

Questions:
1. What made the paper burn?
2. Why would the paper not burn without the magnifying glass?
3. What would it depend on how high we have to hold the lens to focus it?
4. What else beside thin paper can we use to burn?
5. How are forest fires sometimes started by nature?
6. What did the magnifying glass actually do with the sun rays?

Explanation:
What the magnifying glass or lens does, is concentrating all the sun rays that were falling on it, into one point. This increased the intensity of the heat 100 or may be 200 fold, depending on how large and how strong the lens is. You will find that on clear sunny days it is much easier to start the fire, than it is on hazy or cloudy days. The stronger the lens (more convex) the closer we have to hold the lens to the piece of paper to focus the sun rays.

Dry leaves can easily be substituted for the tissue paper. In nature, forest fires are sometimes caused by concentrated sun rays. Lingering dew drops between plant leaves can act as lenses and the concentrated sun rays may fall on very dry leaves, and a forest fire is in the making!

ENERGY POTENTIAL VS KINETIC
THE PENDULUM

WILL THE HEAVY BRICK HIT YOUR NOSE?

Materials: 1. A heavy masonry brick or large bowling ball.   2. A strong nylon rope.

Procedure:
1. Tie the brick or bowling ball to one end of the rope and suspend the pendulum about 3 m from the wall to the ceiling.
2. Before securing the knot in the rope, adjust the length of the pendulum such that the highest swing just reaches your nose when you stand with your back against the wall (see Sketch on right).
3. Make sure that the knots are tight and secure and cannot slip.
4. Now bring the pendulum as close as possible to your nose and let go.
   (CAUTION: DO NOT PUSH THE PENDULUM AWAY FROM YOU!).
5. Stay perfectly still against the wall and let the pendulum swing back.
   (Ask the students: "Who dares to do this?")

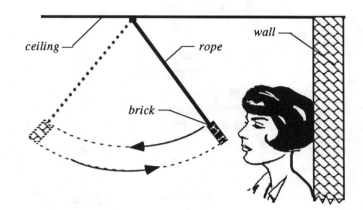

Questions:
1. Why is it perfectly safe to stand against the wall and not get hit?
2. Will the pendulum ever swing higher than its first position?
3. What would give the brick a higher swing?
4. What forms of energy are involved in a swinging pendulum?
5. Where does the energy go? What does it turn into?
6. In which positions of the swing does the pendulum have the greatest kinetic energy? The greatest potential energy?
7. What would happen if the brick was swung in a vacuum?
8. What makes a clock pendulum swing? When does it stop?

Explanation:
When the brick pendulum is released for the first time, it loses some of its energy due to friction at the suspension point and mostly in air resistance. This energy is turned into heat or actually faster moving rope and air molecules. This loss occurs with every swing of the brick and thus each swing becomes lower and lower *(amplitude* gets smaller). A higher swing would only be obtained if we add energy to the pendulum by giving it a push *(do **not** do this while you are still standing against the wall!).*

If this brick pendulum was swung in a vacuum, energy loss would only occur in the rope and the suspension point, and the swinging would last for a much longer time before it stops.

In a clock pendulum the swinging is kept going by the spring of the clock. A small amount of energy is added to each swing by the potential energy of the wound up spring. When the spring is completely unwound, the clock and the swing of its pendulum stops.

ENERGY   ENERGY TRANSFER
                                                      THE PENDULUM

THE TWIN PENDULUM

Materials: 1. Two identical washers (or any other object to make a bob).
             2. Thin string or strong thread.   3. Two vertical stands.

Procedure:

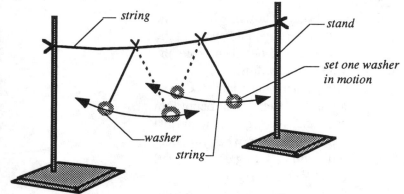

1. Place the two vertical stands about 30 cm apart from each other and tie a piece of string horizontally from one stand to the other.
2. Tie a short piece of string (about 15 cm) to each of the washers (or pendulum bobs) and tie the other end to the horizontal string, about 5 cm apart from each other - make sure that the bobs hang down from the horizontal string at the same height!
3. Let the two pendulums hang perfectly still: now you are ready for the demonstration. Ask: "What will happen to the second pendulum if I start swinging the first one?" Anticipated answer: "Nothing!"
4. Pull out the first pendulum carefully by the bob (hold the string tight) and let go of the bob: OBSERVE!

Questions:
1. What happened to the second pendulum?
2. Where did all the energy of the first pendulum go to?
3. What made the second pendulum pick up the energy?
4. Would a pendulum that was a little longer or shorter than the first one also be able to pick up the energy?
5. Would the second pendulum also pick up energy from the first if they were both tied to a solid metal rod instead of a string?
6. Did the first pendulum completely stop after the energy was transferred?
7. If compared to just one pendulum swinging from a similar support, which of the two systems would swing the longest? Which are the variables?

Explanation:

   Only when the two pendulums are perfectly identical in length, will the energy of the first pendulum be transferred totally to the second. The first pendulum then stops and in turn picks up the energy from the second, etc. until the energy is dissipated into heat, and both pendulums stop swinging.

   The energy from the first pendulum is transferred to the second only because of the flexible horizontal support, which moves in phase with the swing of the second pendulum. These movements keep on strengthening the swing just like the push that we give to a child on a real swing in the park (see also Events 12.8 & 12.7 of ISI-2nd Ed).

Invitations to Science Inquiry – Supplement – Page 80

HEAT   CONDUCTION
                                                        KINDLING POINT

THE BROKEN FLAME

Materials: 1. A Bunsen burner or large candle.
2. Wire gauze (without asbestos) or use the clear part
3. Pair of tongs or tripod.

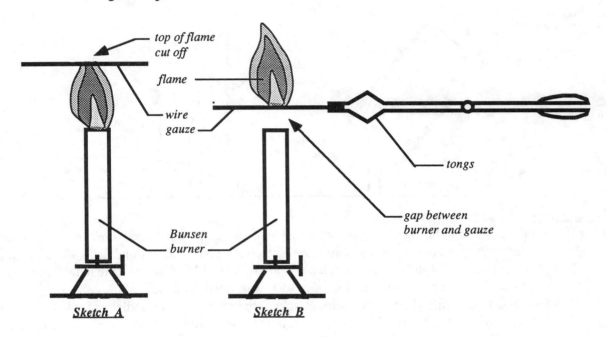

Procedure:
1. Light the Bunsen burner (leave the flame yellow) or light the large candle.
2. Hold the wire gauze (with a pair of pliers or tongs) in the middle of the flame (top of flame is cut off - see Sketch A).
3. Vary the height of the gauze above the burner or candle - bring the gauze slowly down - until suddenly the flame appears above the gauze.
4. Now raise the gauze slowly and observe the gap between the burner and the gauze! - see Sketch B. What happens if you keep on raising the gauze?

Questions:
1. After holding the gauze in the flame, does the flame immediately appear above the gauze? Why? Why not?
2. How far above the burner (or candle) can the gauze be held in order for the flame still to appear?
3. Do you see the flame below the gauze? Before or after the flame appears above the gauze?
4. What makes the flame disappear below the gauze?

Explanation: In the beginning of the demonstration when the gauze was held over the flame, the flame was cut off exactly where the gauze was - Sketch A. This happened because the heat was absorbed and *conducted* away by the gauze, the temperature above the gauze was thus too low to ignite the flame (below *kindling temperature*). Unburned gases (or candle wax vapors) were still present above the gauze, however. After holding the gauze in the flame for a longer time, the gauze is reaching kindling temperature of the gases and the flame above the gauze is suddenly ignited. After raising the gauze slightly above the burner, a gap (no flame) is observed between the burner and the gauze. This happens because all the heat is absorbed by the gauze and also cool gas is being supplied under the gauze, which keeps it below *kindling temperature*.

HEAT

RADIATION ABSORPTION

WHICH COIN WILL STAY ON LONGER?

Materials: 1. An empty juice can (or any other large tin can) of which the top and bottom has been taken off (remove with can opener).   2. A medium size candle, two identical coins.

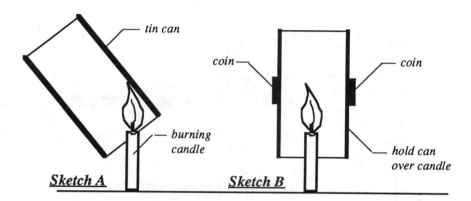

Sketch A   Sketch B

Procedure:
1. Light the candle and blacken one half of the inside of the can with soot from the candle, by holding it sideways close to the candle flame (see Sketch A). Once this is done, let stand and cool.
2. Against the outside of the tin can, attach one coin on each side of the can (the blank side and the soot-blackened side) with one drop of molten wax from the candle.
3. Now place the tin can (with the two coins attached) over the burning candle, and ask the students: "Which of the two coins will stay sticking to the can longer?" - Make sure that you place the can such that the candle is exactly in the center of the can. (see Sketch B).

Questions:
1. What is the cause of one coin dropping first?
2. Which of the coins dropped first?
3. Will the other coin also drop off eventually or not?
4. Would blackening the outside of the can have the same effect?
5. How did the heat travel from the flame to the can?

Explanation:
The coin attached to the side of the tin can that was blackened with the candle soot would fall off first. This is because the heat that was *radiated* from the candle flame was *absorbed* more by the black surface, as compared to the shiny metallic surface. Because of the higher degree of absorption of heat by the black half, the temperature increased more rapidly and thus the wax melted sooner and the coin would drop off before the other one.

If the outside of the can was blackened with soot instead of the inside surface, there would be no difference in the absorption of heat from the candle flame, as the inside surface would reflect the heat rays away in the same fashion. Absorption of heat from outside the can might then have an influence on making the difference in temperature, and thus in the length of adherance of the coins to the tin can.

MAGNETISM   MAGNETIC LINES
                                                                      OF FORCE

THE MYSTERIOUSLY MOVING NEEDLE

Materials: 1. One strong bar magnet, a medium size sewing needle.
2. A wide and shallow glass or plastic tray (to hold water).
3. A small cork or small piece of styrofoam.

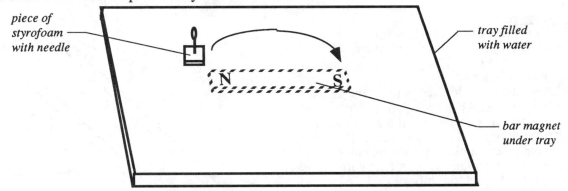

Procedure:
1. Magnetize the needle by rubbing the North pole of the rod magnet several times from the point towards the hole end (see step 1 under procedure for **Make a Needle Compass**).
2. Fill the tray with water (about 2 cm deep) and place it on top of the bar magnet (propping the sides so the water level stays horizontal).
3. Pierce the needle through half of a small cork or a small piece of styrofoam, such that the sharp end points vertically down.
4. Place this needle near the North pole of the magnet and observe! (make sure that the sharp point just floats 2-3 mm above the bottom of the tray. If it is not, just add some more water to the tray).

Questions:
1. What did you observe the needle doing?
2. What would it do if the needle was placed near the South pole?
3. Would rubbing the needle with the South pole of the magnet make any difference in movement?
4. What difference would it make if the needle was rubbed with the North pole but in the opposite direction (from hole to sharp end)?
5. Why doesn't the needle move in a straight line towards the poles?

Explanation:
By rubbing the needle with the North pole of the strong magnet from the sharp end towards the hole end, the needle itself will become a magnet, the sharp end being the North pole and the other end the South. After placing the needle vertically on the water surface, it is free to move, and it is thus repelled by the North pole and attracted by the South pole of the magnet (which is lying underneath the tray). It travels in a curved line following the magnetic lines of the strong magnet. The strongest field is closest to the poles of the magnet and thus the movement of the needle is fastest when approaching each of the poles.

A needle that is not magnetized would just move straight towards whatever pole is closest to the needle. What would happen with the needle if we had a horseshoe magnet under the tray of water?

MAGNETISM   ELECTROMAGNETIC INDUCTION

THE DROPPING RACE

Materials: 1. Two 3 ft (1 m) long brass tubing, 1/2" (1.3 cm) in diameter.
2. A cow magnet (with a diameter slightly smaller than the tube)
3. Other small cylinders similarly shaped as the cow magnet, like an A-battery or a cap of a marker, etc.
4. An aluminum pie plate or metallic jar lid.

Procedure:

1. Give one of the brass tubes to a volunteer (student) and let him choose any of the objects to race with you, who has the cow magnet.
2. The object of the race is to let the object stay in the tube as long as possible, so that a later click on the metal plate indicates the winner. Tell the students that they may change anything (like: dimensions, weight, etc) of the object, as long as it will fall freely through the tube.
3. Hold the brass tube almost vertically above the metal plate and get ready to let the cylindrical object slide into the tube. Let the student do the same.
4. Tell the students: "On the count of three, let go of the object! Listen to the clicks!

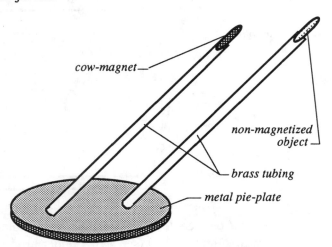

Questions:

1. What are the variables that are working on this race?
2. Which is the manipulated variable? (the variable that we chose to change?)
3. Which is the responding variable? (the variable that we measured?)
4. What made the cow magnet always win from the other objects?
5. If the brass tube was a coil of wire, what would the falling magnet induce in the wire?
6. What would a circular current in the tube change the tube into?
7. Would flipping the polarity of the magnet have any influence on the falling rate?
8. What variable can we change to let the magnet fall even slower?

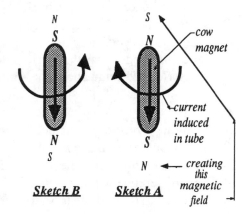

Explanation: The cause for the cow magnet to slow down in the falling rate is the *induction* of a current or electron flow in the tubing perpendicular to the direction of the fall. The falling magnet causes a change of the magnetic field, which therefore induces a current in the tube (just like in a coil of wire). These currents turn the tube into a magnet with opposite poles, only at the place where this change is taking place. The falling cow magnet is therefore slightly held back in the fall.(see Sketch A) When the magnet is turned upside down and then released, the opposite current is induced and an opposite field is created, also holding back the magnet in its movement.(see Sketch B).

The variables that are working on this particular race are: tube dimensions, magnet dimensions, cylindrical object dimensions, slant of the tube, moment of release, etc. The manipulated one here is of course whether the falling object is magnetized or not, and the responding one is the time the object took to come out of the tube. This slowing down of the falling rate can be increased by 1. increasing the strength of the magnet, 2. slanting the tube more, 3. lengthening the brass tubing, etc., etc.

| STATIC ELECTRICITY | ATTRACTION OF UNCHARGED OBJECTS |

THE ELECTRIC METER STICK

Materials: 1. A wooden meter stick. 2. A plastic comb, a piece of flannel or wool.

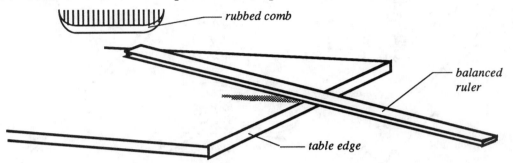

Procedure:
1. Find a table with a rather sharp edge and let half of the ruler protrude over the edge of the table.
2. Tip the ruler down over the edge, and by pulling it more or less over the edge, let the ruler balance in a slanted position. (see Sketch above).
3. Rub the piece of flannel over the comb and approach the higher end of the ruler with the charged comb.
4. Now touch the upper end of the ruler with the charged comb and wait a bit till the ruler stops moving.
5. Rub the comb again and approach the same end of the ruler that was touched before. Which way does the ruler move now?

Questions:
1. Which way did the upper end of the ruler move when first approached with the charged comb?
2. Was the ruler end attracted or repelled by the comb?
3. After touching the same end of the ruler with the charged comb, was the ruler attracted or repelled by the comb?
4. If the comb was positively charged, can you explain in terms of moving electrons why the ruler behaved the way it did before it was charged? After it was charged (touched by the comb)?

Explanation:

The wooden ruler balances on the edge of the table on its center of gravity and becomes a very sensitive balance. A minute force will tip the ruler up or down on one side.

By rubbing the comb with a piece of flannel, it gets charged with static electricity (most likely positive). Approaching the uncharged ruler with the positive charge attracts the negative charges in the ruler towards the comb. As negative is attracted by positive, the ruler tips up (Sketch A). Touching the ruler with the comb withdraws the negative charges out of the ruler and leaves it positively charged. The second time we approach the ruler with the comb, the ruler moves down (see Sketch B).

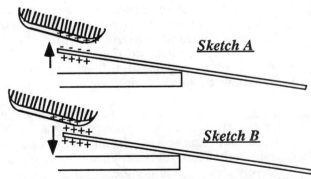

Sketch A

Sketch B

Invitations to Science Inquiry – Supplement – Page 85

ELECTRICITY — STATIC CHARGES

THE LEVITATION ACT

Materials: 1. A hard plastic pipe (PVC) 3/4" x 3 ft long. 2. Two pieces of foam rubber (1x5x5").
3. A thin strip of flexible styrofoam (from wrapping material).
4. A small cork to fit in the pipe.   5. A toothpick.

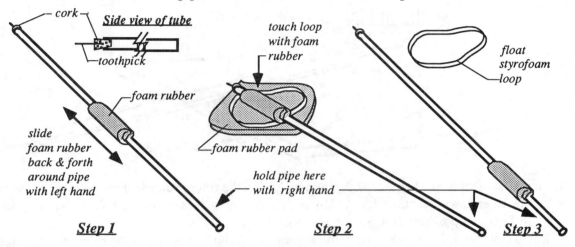

Step 1   *Step 2*   *Step 3*

Procedure:
1. Fit the cork in one end of the pipe and push the toothpick end into the center of the cork (see sideview of tube, part of *Step 1*). Wrap one piece of foam rubber around the pipe and tape it.
2. Make a continuous loop out of the thin strip of styrofoam by taping the ends together, and place it on the other foam rubber piece.
3. Hold the pipe in one hand and rub it back and forth with the foam rubber in the other hand.
4. Leave the foam rubber at the cork end of the pipe and touch (roll) the foam rubber to the styrofoam strip (see *Step 2*), then slide the foam rubber back to where the right hand holds the pipe.
6. Now pick up the styrofoam strip with the end of the toothpick and throw the strip in the air. Immediately move the pipe below the slowly falling styrofoam strip and balance it above the pipe by changing the angle and/or moving it sideways. (see *Step 3*).

Questions:
1. What is the function of the foam rubber wrapped around the PVC pipe?
2. What do we need the other foam rubber piece for?
3. What do we create on the pipe by rubbing it with the foam rubber?
4. What force kept the styrofoam strip floating above the PVC pipe?
5. If we call the charge created on the pipe positive, what charge is created on the foam rubber?
6. Why do we charge the styrofoam strip with the foam rubber and not with the pipe?
7. When approaching the styrofoam strip with the positively charged pipe, how do the charges shift in the styrofoam strip?
8. Why can't we pick the styrofoam strip up with our fingers?
9. What other materials could we use instead of the styrofoam strip?

Explanation: Since electrons or negative charges are more movable, we may assume that a positive charge is left on the pipe by the rubbing action. The inside of the foam rubber is thus negatively charged, and because it is an *insulator, -a substance where electric charges cannot flow easily* - the outside of the foam rubber is positively charged. When we approach the styrofoam strip with this positively charged foam rubber, all the negative charges in the strip are attracted to the surface and when touching occurs, transfer of electrons takes place and the strip is left temporarily positively charged. Now that both pipe and strip are positively charged, we can balance the strip above the pipe and the *repulsion force* will keep it suspended above the pipe, provided the strip is light enough in weight. A very light feather or other insulating material may replace this styrofoam strip.

Picking up the strip with our hands will neutralize the strip since we are conductors of electrons and thus it is important to pick the strip up with the tip of the toothpick.

CURRENT ELECTRICITY CIRCUITS

THE BALLOON FUSE

Materials: 1. A small balloon, aluminum foil, simple switch.
2. A strong battery (6V) or five 1.2 V D-batteries.
3. Bulb (6V), bulb holder, copper wire, steelwool.

Procedure:
1. Cut two strips of aluminum foil and tape one strand of steelwool between the two strips against the balloon (see Sketch).

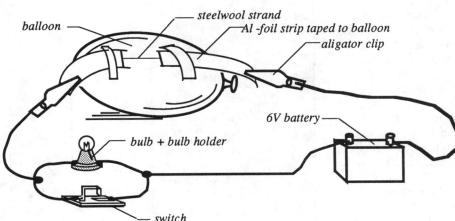

2. Use the copper wire to complete the circuit as laid out in the Sketch. (same circuit as in Event 10.6 of ISI- 2nd Ed)).
3. Show students that the light is burning because the circuit is complete.
4. Draw their attention now to the switch, which will make the short circuit when closed, and the single strand of steelwool on top of the balloon.
5. Insist on everybody's attention, then push the switch: BANG! The fuse broke and the light is off.

Questions:
1. What made the balloon pop when the switch was pushed?
2. What is a short circuit?
3. Which way would electricity rather flow, through the switch or the bulb?
4. Why did the light go out after the balloon popped?

Explanation:
The single strand of steelwool had the function of the fuse wire. When the circuit was shorted by pushing the switch, a sudden surge of electric current was flowing through the circuit causing the single strand of steelwool to heat up. As this strand of steelwool was taped against the balloon, it melted the rubber and bursted it. With the steelwool strand broken, the circuit in which the bulb is wired is thus incomplete, and so the light stays out.

In our daily life, in most electrical appliances, all automobiles and homes are protected with fuses. These fuses can let a certain amount of electricity pass through (f.i. 10, 15, or 20 amperes). When a sudden surge of electrical flow occurs the fuse breaks. In the new homes they are called ***circuit breakers*** which click off when a short circuit occurs and can be reset after the short is repaired.

LIGHT   REFLECTION

THE SIMPLE PERISCOPE

Materials: 1. A couple of rectangle mirrors.   2. A large piece of cardboard (1x2 ft)
3. Scissors & masking tape.   4. Protractor & ruler.

Procedure:
1. Lay the rectangle cardboard flat on another piece of cardboard and draw three lines lengthwise on it, so that we can divide it into four equal parts/strips. (See Sketch A).
2. Cut square windows in strip 2 and 4, and 45 degree angle slots in strip 1 and 2 as in Sketch A.
3. Score the cardboard on each of the three lines, so that it can fold into a square tube. Fold the cardboard along the scores and tape one edge with the masking tape.

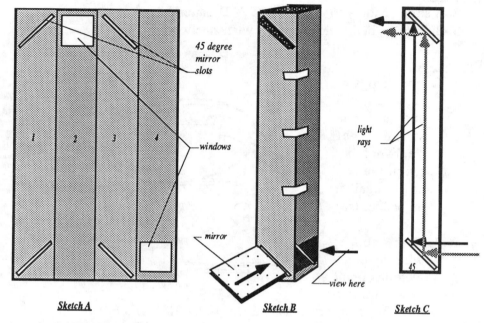

4. Slide the mirrors into the angled slots, such that they face the open windows and tape them in place. By holding the tube horizontal you can look around corners, when holding it upright, you can look over a wall.

Questions:
1. What is the purpose of a periscope?
2. Could we achieve the same objective with one mirror instead of two?
3. Where are periscopes mostly used in our daily life?
4. What shape does the actual periscope usually have?
5. Why is the angle of 45 degrees so critical in building a periscope?
6. Would it be possible to use the periscope the other way around? - to look at things that are much lower than ourselves (lower apartments, coal mines, etc.)?
7. How would you construct a periscope that would look from balcony into another balcony at a lower or higher level? How do the mirrors and windows have to be placed?

Explanation: The purpose of a periscope is generally to look at a point that is much higher than we actually are. The most common use of the periscope is in submarines. In order to look above the water while we are in the submarine (under water), we need the periscope. Since the incoming ray makes the same and equal angle with the normal *(angle of incidence)* as the reflected ray with the normal *(angle of reflection)*, it takes exactly 45 degrees to make the light to reflect at a right angle (90 degrees). This is why we need to place the two mirrors exactly at a 45 degree angle. It is not possible to use only one mirror and achieve the same objective as the periscope, because one mirror will turn the image upside down. (see sketch)

In order to look from balcony to balcony at a higher or lower level, we need an additional mirror so that we can see things right side up. A periscope can certainly be used to look at things that are lower than us, like in mines. All we need to do is use the top mirror to look into the same periscope!

LIGHT   REFRACTION

SPEAR-FISHING, ANYONE?

Materials: 1. A small empty aquarium, or clear deep plastic tray.
2. A cardboard cylinder (core of a paper towel).
3. Two wire hangers, a piece of cardboard.
4. A washer or nut, masking tape.

Procedure:
1. Fill the aquarium or clear, deep, plastic tray almost full with water.
2. Hang the washer on the coat hanger hook and bend the hanger such that it will hang over the edges of the aquarium (the washer is thus hanging in the water).
3. On the other end of the container, tape a piece of cardboard, such that it hinges around the top edge; then tape the cardboard cylinder on the cardboard.

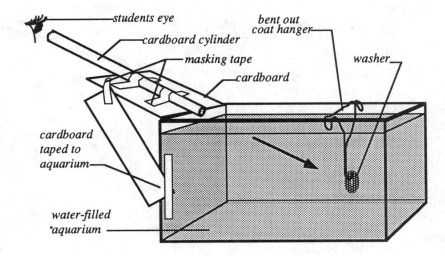

4. Have a student look with one eye through the cylinder and direct it towards the washer. Once it is sighted, fix the angle of the cylinder by taping another piece of cardboard against the cylinder and the aquarium.
5. Now push a long straight wire (straightened from the other coat hanger) through the cardboard cylinder: Does the "spear" hit the "fish"?

Questions:
1. Where does the point of the "spear" end up in the aquarium relative to the "fish"?
2. Why will it never hit the "fish", when aimed directly at it and at an angle with the water surface?
3. When only (at what angle with the water surface) will the "spear" hit the "fish" right on?
4. Where do we have to aim the spear relative to the fish when we go spear-fishing?
5. How can we eliminate the error (and hit the fish) when going spear-fishing?

Explanation: The light ray from the washer (or fish) under water is refracted, passing from water into air. Since air has a smaller refraction index compared to water, the light ray is bent away from the normal. Starting from the observing eye, the light ray is going from air into water: from less dense to denser medium, thus the light ray is bent towards the normal. This is the reason why swimming pools usually look much more shallow than they actually are.

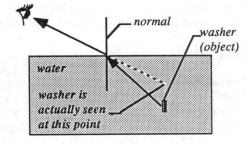

In order to hit the fish with a spear, we therefore have to throw the spear lower then where we see the fish, because actually, it is swimming much lower then where we see it. Only when we are looking straight down (at 90 degrees with the water surface, along the normal) into the water, can we throw the spear right at the fish where we see it, and actually hit it too!

Invitations to Science Inquiry – Supplement

LIGHT   REFRACTION

THE MAGIC OIL

Materials:
1. A clear drinking glass.
2. 2 small Pyrex test tubes.
3. Frying oil (preferably Wesson)
4. A pair of tweezers
5. A small piece of fine emery cloth

Procedure:
1. *Preparation:* a. Pour the oil into the drinking glass until it's full.
   b. Take one of the test tubes, wrap it in a piece of paper, and hit it with a hammer (or other hard object) so that it will break in pieces.
   c. Take the other test tube and scrape away all of the pyrex brand on the tube with the fine sand paper.
   d. Place the whole test tube in the oil-filled glass, making sure that no air bubbles are left in it. Now you are ready for the demo.

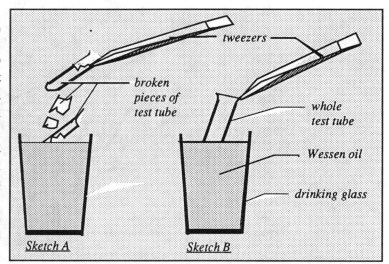

2. *Demonstration:* a. Tell students that you have a glass full of magic oil, and that it can work wonders.
   b. Show the broken pieces of glass tubing and plunk them in the glass of oil (leave the piece(s) with the PYREX mark out).
   c. With the pair of tweezers in your hand (say abacadabra!) and slowly take the whole test tube out of the oil!

Questions:
1. Why did we have to take the PYREX mark off the test tube?
2. What did you observe when the glass pieces were put in the oil?
3. What is the reason that we could not see the PYREX glass pieces or the whole test tube, once they were put in the oil?
4. What would happen if a bubble of air was left in the test tube?
5. Would this trick also work with other brands of glass or oil?

Explanation:

PYREX glass and Wesson oil happen to have *refractive indexes* that are very close to each other. This is the reason why the Pyrex pieces of glass were almost invisible in the oil. The light rays that pass through the oil would pass through the Pyrex glass with the same speed, resulting in no refraction when passing from oil into glass or vice versa. In other words, the light rays pass through the oil and the glass in a straight line. Store windows are made of plate glass that have refractive indexes so close to that of air, that we sometimes bump into them thinking that it was a doorway!

If an air bubble is trapped in the test tube, it will definitely be seen in the oil. Also if the PYREX label on the test tube is not removed, a floating label will be seen in the oil. It is therefore necessary to remove all labels and not leave scratches on the glass.

LIGHT   REFRACTION
                                                                     TOTAL REFLECTION

WHY DO WE SEE TWO COINS?

Materials: 1. A penny or nickel (or any other coin).
2. A clear colorless drink glass, or beaker.

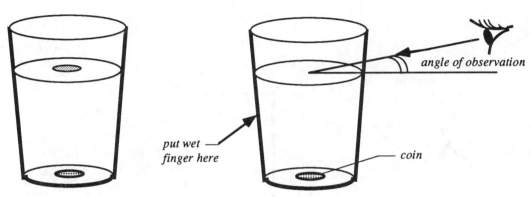

Procedure:

1. Fill the glass about three quarters full with water.
2. Drop the coin in the glass and maneuver the glass till the coin sits in the middle of the bottom.
3. Let the glass with the water and coin now stand still on a table and observe the surface of the water from about a 15-20 degree angle.
4. Touch the far side of the glass with a dry finger: does anything happen to the image of the coin?
5. Wet your finger, then touch the far side of the glass with the wet finger: what happens to the image of the coin?

Questions:

1. What makes us see another coin on the surface of the water?
2. Does the image of the coin get inverted?
3. What do we see when the angle of observation is larger?
4. Why is the image of the coin disappearing when the glass wall is touched with a wet finger?
5. Where is reflection occurring, and where refraction?

Explanation:

There is **total reflection** of the image of the coin occurring at the back wall of the glass, then **refraction** occurs at the surface of the water.

If a wet finger is placed on the back wall where the total reflection is supposed to occur, a layer of water is actually placed on the outside of the glass. This equalizes the refraction indexes on both sides of the glass (because they are of the same medium: water), and total reflection is thus eliminated.

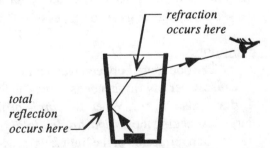

If a dry finger is held at the same spot, it will not be seen, because on the outside of the glass is air rather then water. When we look from a denser medium (water) to a less dense medium (air) at a particular angle, total reflection occurs, and thus the finger is not seen.

Invitations to Science Inquiry – Supplement – Page 91

LIGHT

REFRACTION
TOTAL REFLECTION

USE WATER AS A MIRROR?

Materials: 1. A small candle & matches. 2. A 400 ml beaker or regular large drinking glass.

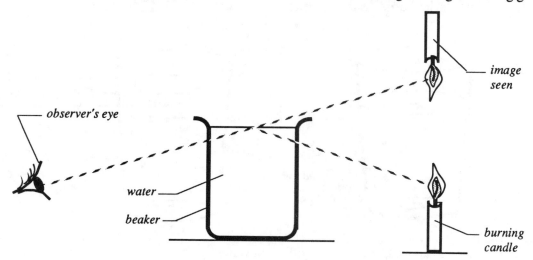

Procedure:
1. Light the candle and place it on a shelf or stand at about eye level in a dark corner of the room (when sitting it should be somewhat lower but still at eye level).
2. Fill the beaker or glass full with water.
3. Hold the beaker a little higher than the candle, and look up against the water surface, such that the candle flame will reflect in it - the height of the beaker may need adjusting in order to see the reflection of the candle flame (see Sketch above).

Questions:
1. Why do we see an inverted candle flame image?
2. How can water give a reflection like a mirror?
3. Why do we need a dark background to see the reflection?
4. What other substances will act similarly like the water?
5. What property was different between the water and the air?
6. Would different gases or gases of different densities be able to act like a mirror?

Explanation:
The boundary between water and air act like a mirror because water has a higher density than air and its *refraction index* is also higher. When the *angle of incidence* is too large (exceeding the *critical angle*), *total reflection* occurs rather than refraction of the light rays (see Sketch on right). Hot air bordering on cold air will act similarly. Hot air over a road on a hot summer day will give one the illusion of water on the road. Mirages over a desert are formed in the same way.

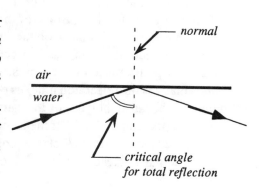

Invitations to Science Inquiry – Supplement –

LIGHT   REFRACTION
                                                            TOTAL REFLECTION

THE REFLECTING BRICK WALL

Materials: 1. A long (minimum 10 m) wall on a hot sunny day.
2. A bright metallic object (like a shiny key).

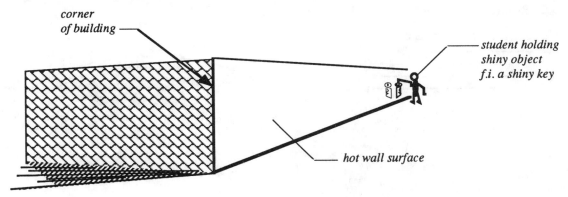

Procedure:
1. Take the class outdoors on a bright sunny day. Stand at one end in front of a brick or concrete wall against which the sun is shining.
2. Let one of the students hold a shiny metallic object, like a key or knife, about 5 to 10 cm from the wall, such that you can see it.
3. Observe an image of the object in the wall! (If you don't see the image, move in closer to the wall or let the student move the object closer to the wall).

Questions:
1. What made the wall act like a mirror?
2. Why does it have to be a sunny day, where the sun shines against the wall?
3. What other surfaces would act in the same manner?
4. In what other situations may we encounter the same phenomena?
5. Why would a shiny object work better than a dull object?
6. Why can we only see a reflection when we are so close to the wall?

Explanation:
When the sun shines against the wall, it heats the wall up and the hot wall heats the air surrounding it. This hot layer of air has a different *index of refraction* as compared to cooler air. It therefore acts as a different medium against which light can be totally reflected. This angle of total reflection occurs only at very large angles of incidence. This is why we have to be so close to the wall to see the reflection of the shiny object. A shiny object works better because it can easily be seen from such a distance.

This same phenomena occurs in the mirages that we see over deserts. The reflecting medium in this case is colder air, which hovers above the hot air over the desert sands. Another example is when we are driving on a hot day, we may see far off on the road a shiny spot as if it was wet or oily, but on approaching it, it suddenly disappears. Here the hot air above the road acted like a mirror, giving total reflection only at large angles of incidence.

LIGHT   REFRACTION & SCATTERING
                                         TOTAL REFLECTION

MAKE YOUR OWN RAINBOW

Materials: 1. A garden hose + sprayer nozzle (connected to water outlet).
2. Bright sunlight (when sun is low in the sky).

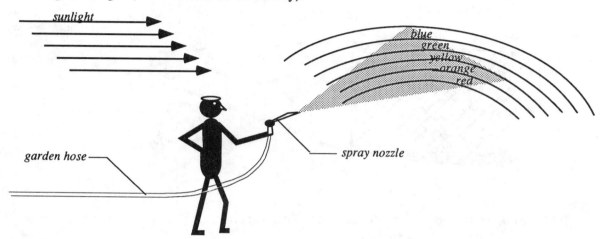

Procedure:
1. Take the class outdoors on a bright sunny day. Attach your garden hose to the outdoor tap, hold the spray nozzle in your hand and position yourself such that you face away from the sun (the sun behind you).
2. Squeeze the spray nozzle and produce a fine spray of water in front of you, and Voila! A beautiful rainbow!

Questions:
1. What are the conditions for seeing a rainbow?
2. When is a rainbow actually seen in nature?
3. What makes the different colors in the rainbow?
4. What color is on top, what color on the lower side?
5. Why is the rainbow always circular/ a segment of a circle?
6. Why is the rainbow never a square, triangle, or straight line?
7. Why does the sun always have to be behind us to see a rainbow?
8. Where do we see rainbows, besides In the sky?

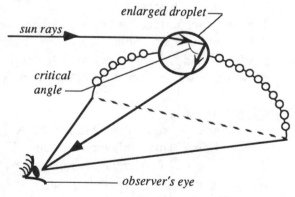

Explanation:

The conditions for seeing a rainbow are: the sun has to be behind us, we have to look away from the sun, and facing the fine droplets in the sky (clouds after a rain). These droplets can be a few meters, a few hundred meters or a few thousand meters or kilometers away from us.

The rainbow is always a segment of a circle, because it is a part of the circular base of a cone. These are the only droplets that we can see because of the *critical angle* that gives a *total reflection* of the sun ray falling into it. The blue color is always on top since it is the most refracted one (with the largest refraction). Other places where we could see rainbows are: fountains, waterfalls in bright sunlight (provided we stand in the right position).

Invitations to Science Inquiry – Supplement – Page 94

SOUND   PITCH

PLUCK A RUBBER BAND

Materials: 1. A small medium thick rubber band.

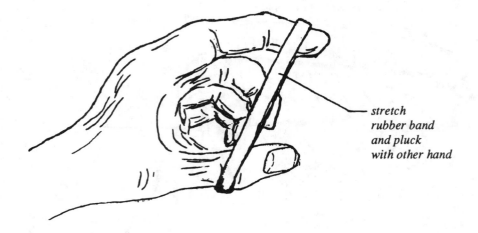

stretch rubber band and pluck with other hand

Procedure:
1. Place the rubber band between thumb and forefinger, and stretch it a little.
2. Hold it close to your ear and pluck it with the other hand.
3. Stretch the rubber band by widening the gap between thumb and forefinger and pluck again. What pitch do you hear now?

Questions:
1. Did the pitch of the sound change after widening the gap?
2. Which properties of the rubber band changed while being stretched?
3. What property changes when a guitar string is tightened?
4. Which properties do not change when a guitar string is tightened?
5. How do these properties compare to those of the rubber band?
6. In what ways can we change the pitch of a guitar string?
7. How could we change the pitch of the rubber band?

Explanation:
When the rubber band stretched and was plucked again, the pitch of the sound was staying about the same, if not getting a little lower. This is definitely contrary to what one would expect, which is a higher pitch for the stretched rubber band.

When a guitar string is tightened, the pitch becomes higher, because the tension is higher, thus the ***string vibrates with a higher frequency***; the length and the density of this string, however, is staying constant.

With the stretching of the rubber band, all three properties: tension, length, and density change. The higher tension tend to increase the pitch, but this is compensated by the increase of the length, which tends to lower the pitch.

When the length of the rubber band is held constant, the pitch changes similarly like a regular guitar string. This can be done by stretching the rubber band over an empty open box.

SOUND   DOPPLER EFFECT

HEAR THE DOPPLER EFFECT

Materials: 1. A small alarm clock (with a bell sound and handle on top)
2. A piece of strong (nylon) string - about 2m (6 ft) long.

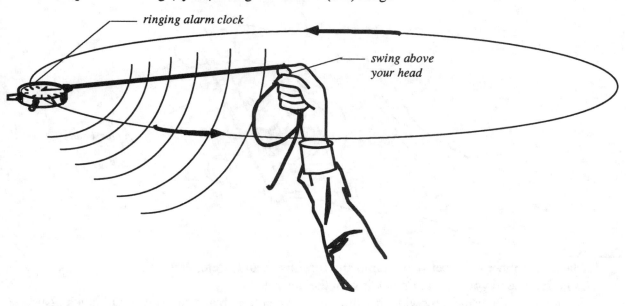

Procedure:
1. Tie the end of the string to the top handle of the alarm clock.
2. Set the alarm off, grab the other end of the rope and swing the alarm clock above your head. (Make sure you have enough room to swing the clock around)
3. Listen carefully to the pitch of the alarm clock. Does it stay the same?

Questions:
1. Why did we hear two different pitches of the bell?
2. Was the clock moving away or coming towards us when we heard the higher pitch?
3. What would we hear if we swung the clock faster around?
4. How would the pitch change if we used a longer string?
5. What examples can you name in daily life where this particular effect can be experienced?

Explanation: The pitch of a tone that we hear is determined by the *frequency* or number of waves that reaches our ear drum. The more waves or the higher the frequency of sound, the higher the pitch. When the alarm clock is swung above our head, it moves away or comes towards us. When it comes towards our ear drum, more waves reaches it so that we will hear a higher pitch. When the clock moves away from our ear, fewer waves reaches our ear, and so we hear a lower pitch.

This principle was discovered by **Christian Johann Doppler** in 1842. It is being used in astronomy to determine how fast the earth is rotating around the sun among others. In the field of radar, it makes it possible to do speed checks on the highways. The *Doppler Effect* can be experienced when driving in a car and another on-coming car passes by while the horn is on continuously. When sitting in a train sometimes we can also hear the higher pitch of another on-coming train's whistle and as soon as it passes by us, we hear a lower pitched whistle. This same effect is heard when a fire siren from a fire station is on: a higher and lower pitch is heard alternatively with a constant period. This happens because of the horn of the siren which rotates on top of the fire station with a constant speed. When the horn turns towards us we hear the higher pitch, and when it turns away from us, we hear the lower pitch.

SOUND   REFLECTION

THE ELLIPTICAL WONDER

Materials: 1. Two thumbtacks, a pencil, and a marble.   2. A thick thread or thin rope.
3. A smooth table surface or large cardboard. 4. A long 3 cm wide cardboard strip and molding clay.

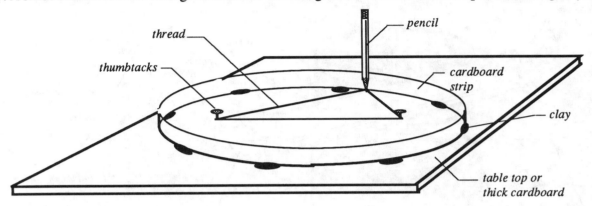

Procedure:
1. Press two thumbtacks about 30 cm apart in the cardboard or table top (this can be closer or wider apart depending on the size of the surface area of the table or cardboard).
2. Make a loop with the thread, where the total length of the thread is 5-10 cm longer than twice the distance between the thumbtacks.
3. Hold this loop of thread tight between the thumbtacks and the pencil tip (holding the pencil vertically) and describe a perfect ellipse with the pencil on the cardboard.
4. Build a wall with the cardboard strip and the molding clay on this ellipse by pressing blobs of clay between the strip and the table (or cardboard).
5. Roll a marble from one focal point (thumbtack) In any direction to the elliptical wall. Which way does it bounce back?

Questions:
1. Why does the marble always bounce back to the other thumbtack?
2. What points are the thumbtacks in the ellipse?
3. How large an elipse can be constructed this way?
4. How would sound or light waves bounce off the elliptical wall?

Explanation:
The reason why a perfect ellipse results by keeping the thread tight between the two focal points and the pencil is, because the sum of the two radii stays the same.

Another feature of the ellipse is that the ***angle of incidence*** (i) of one radius is the same as the ***angle of reflection*** (r) of the other radius (see Sketch on right). This makes it possible that softly spoken words in one focal point can be heard at another focal point quite a long distance away from the first focal point.

Invitations to Science Inquiry – Supplement – Page 97

FORCES
TORQUES
CENTER OF GRAVITY

THE BURNING CANDLE SEESAW

Materials: 1. A long, cylindrical candle & matches. 2. Two identical tin cans.
3. Two narrow, long nails or pins   4. Two 5x8" cards.

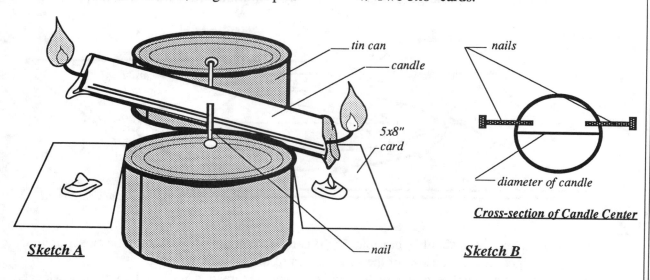

Sketch A

Cross-section of Candle Center

Sketch B

Procedure:
1. Carve away some of the wax from the bottom end of the candle and let the wick stick out.
2. Let the candle balance on your finger to find the center of gravity. Once this point is found, make a small scratch in the candle and push the two nails in each side of the candle, such that they are lining up in each others extension, slightly above the diameter of the candle (Sketch B).
3. Place the two cans next to each other with a space in between, equal to slightly more than the thickness of the candle.
4. Rest the nails (which are already in the candle) on the rims of the cans to make the seesaw.
5. Make sure that the candle balances to a horizontal position by scraping wax off the heavier end if needed, then place the 5x8" cards under each end of the candle.
6. Light both ends of the candle and start the seesaw by pushing one end of the candle down.

Questions:
1. What is the function of the nails?
2. Why do we have to make sure that the candle is in perfect balance before lighting it?
3. What keeps the candle rocking back and forth? - What keeps the seesaw going?
4. What will happen if we <u>don't</u> push one side down to start the rocking?
5. What will happen if - after balancing the candle on the nails - we only light one end?
6. Could we start the seesaw without scraping the wax of the heavy end of the candle? How?
7. How long will the seesawing continue? Or will it?
8. Why do the nails have to be placed slightly higher than the diameter of the candle?
9. How does this candle seesaw differ from a real seesaw with people on it?
10. Where are the forces working and what kind of forces are working on the seesaw?

Explanation: The function of the nails is to work as the *pivot* of the seesaw. This has to be placed slightly higher than the diameter, so that the candle's *center of gravity* stays below the pivot point. If this center of gravity is too high, there is a danger that the candle will make a complete turn and flip over. This candle seesaw works, because it keeps on loosing one drop of candle wax weight at each end and its doing it alternatively. When one end looses one drop, the other end becomes heavier and thus goes down because of a *torque* or *moment* working on this end. The difference with the real seesaw with people, is that no weightloss is occurring in the latter, only a slight upward force with the legs on each end of the seesaw makes it go.

FORCES   CENTER OF GRAVITY

WHERE IS THE BALANCING POINT?

Materials:
1. Pieces of cardboard.
2. A few head pins or tag pins.
3. A stone or weight.

2. A pair of scissors, and a nail..
3. A thin string or thread.
4. Pencil and foot ruler, and corkboard.

Procedure:
1. Cut an irregular shape out of the cardboard piece (like in the sketch above).
2. Make a hole at one end of the cardboard piece with a nail and let the cardboard hang freely from a pin which is pushed in the corkboard.
3. Now tie a string or thin thread to the pin and on the other end of the string tie a stone for weight (plumb-bob) and let it hang freely from the pin.
4. Place the ruler next to the string and draw a pencil line on the cardboard.
5. Take the pin out of the corkboard and turn the cardboard piece about 90 degrees or more and repeat points 2, 3, and 4.
6. The intersection of the two lines on the cardboard piece is the center of gravity. Show that it can actually balance on the tip of your finger, if placed exactly at this intersection.

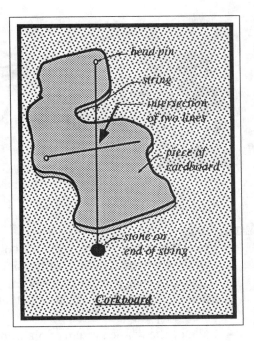

Questions:
1. Why does the cardboard have to hang freely from the pin?
2. Why do we make a hole in the cardboard with a nail first before hanging it from the pin?
3. Would a third line be necessary to determine the balancing point?
4. Where would a third or fourth line definitely go through?
5. How can we define the center of gravity? Or balancing point?
6. Is this point always on the object itself? Or can it be outside the object?
7. How would the object look like, if its center of gravity is located outside itself?

Explanation: The definition of *center of gravity* of an object is *a point in which you can think all the mass is concentrated*. This is why the center of gravity is definitely on the plumb-line that hangs from any point of the object. In order to find it, all we need is two of these plumb-lines, and the intersection of these two lines will therefore be the center of gravity, because only this point is common to the two lines. A third or fourth plumb-line is therefore not necessary, because these lines will also go through the center of gravity.

This **center of gravity** is sometimes also called *the balancing point*, because when we support the object at that point, for example with our finger tip, it will balance. It is not necessary that this point is located on the object itself. It is entirely possible that this point is outside the object. An example would be that of a boomerang (see Sketch on right). We would not be able to draw the center of gravity on the object in this case. Can you think of other objects, where this balance point is outside itself?

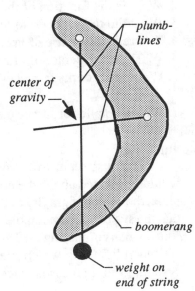

Invitations to Science Inquiry – Supplement – Page 99

FORCES   CENTER OF GRAVITY

THE PLATE CAROUSEL

Materials: 1. A regular dining plate and 4 forks.   2. Two corks or a small raw potato.
3. A corked bottle and a needle.

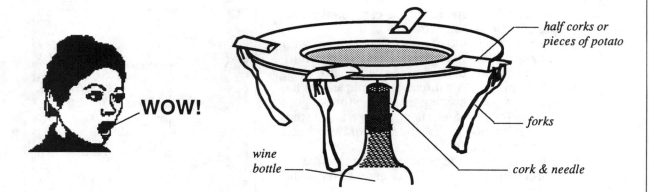

Procedure:
1. Insert the needle vertically in the cork of the bottle and place the bottle in the middle of the table.
2. Cut the two corks in half with a sharp knife (so that the cut surfaces are smooth) or, when corks are not available, cut four equal pieces of potato out of a small raw potato.
3. Stick the four forks in the four half corks (or four chunks of potato) and hang them from the edge of the plate (see Sketch).
4. Balance the whole system of plate and forks on the blunt end of the needle (or the sharp end if you can push the blunt end in the cork).
5. Now gently blow against the forks (or tap against them) in one direction to make the plate rotate.

Questions:
1. Could the plate balance on the needle without the forks?
2. What is the function of the hanging forks?
3. What did the forks do to the centre of gravity of the system?
4. Where is the centre of gravity of the plate located without the forks?
5. Where is the centre of gravity of the system of plate and forks located?
6. Would half corks or pieces of potato be better to make the system more stable, or does it matter at all?
7. Would smaller or larger pieces of potato be better to make a more stable system? Why?

Explanation:
The plate on its own would not be able to balance on the needle, because its *center of gravity* would be located above the point of support (pivot point). By hanging the forks from the edge of the plate, the center of gravity of the system got lowered. It got lowered far enough below the *pivot point* that the system became stable. The more mass is placed below the pivot point, the stabler tne system. Thus, when comparing the half corks with the pieces of potato, the first would be better, as the corks are much lighter in weight (less mass). The bigger the pieces of potato we use, the less stable the system, as more mass is placed above the pivot point - which is the top of the needle.

FORCES — CENTER OF GRAVITY / HUMAN BODY

THE UNREACHABLE CUP

Materials: 1. A cup or pop can (or any other small object).

Procedure:
1. Place a cup or pop can about 20-30 cm in front of your feet on the floor and show how easy it is to pick it up without moving or bending your knees.
2. Let someone (student) stand straight with his/her back against the wall and with his/her heels touching the wall.
3. Place a cup or pop can about 20-30 cm away in front of the student's feet on the floor.
4. Let the student plck up the cup from the floor without bending the knees and without falling forwards. Is it possible?

Questions:
1. What makes it so difficult to pick up the cup from the floor?
2. Why do we tend to fall forward when bending our body to the front?
3. Let the student stand away from the wall and let him/her pick up the cup from the floor without bending the knees. Observe the student from the side. What happens to the legs and lower part of the body?

Explanation:
When picking up an object from the floor without bending our knees, our legs and lower part of our body have to move backwards in order to stay balanced. The *center of gravity* of our body has to remain above our feet, which is the *pivot point,* supporting our body.

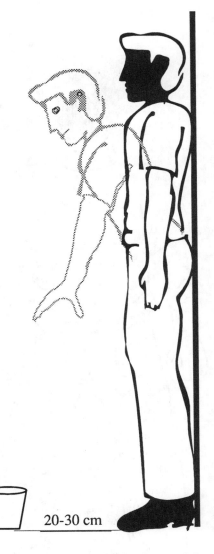

Standing straight with our back against a wall makes it impossible to move the lower part of our body backwards. The forward bending of the upper body shifts the body's center of gravity towards the front of the pivot point (our feet) and the whole body topples or falls forward.

If there is no wall or one that is not too easily accessible, like in an auditorium or the outdoors, you can let two students stand back to back and heel to heel, and let them pick up the cup from the floor in front of each one of them at exactly the same time (on the count of three).

FORCES   CENTER OF GRAVITY
                                                    HUMAN BODY

STUCK TO THE WALL?

Materials: 1. Just yourself and the wall.

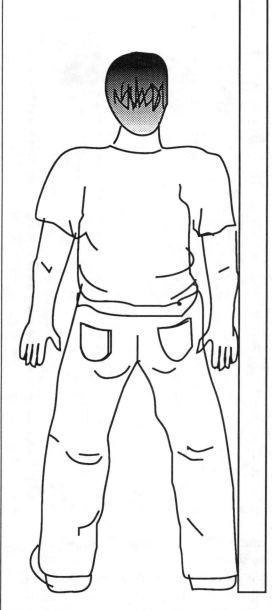

Procedure:
1. Stand up straight with your right foot and your right shoulder aqainst the wall.
2. Try to move your left leg without falling or taking a step.
3. Now turn around, stand straight and touch the wall with your left foot and keep your left shoulder pressed against the wall.
4. Try to lift or move your right leg: IMPOSSIBLE!

Questions:
1. What makes it impossible to move your left leg when your right foot and right shoulder are touching the wall?
2. What did you have to do in order for your right shoulder to press against the wall?
3. What would happen if someone else forced your left leg to move while you are standing in that position?
4. Where is your center of gravity located when you are standing up straight?
5. Where does your center of gravity have to be located in order for you to be able to lift your left leg without falling?

Explanation:
  In order to stand straight with your right foot and right shoulder touching the wall, your left leg has to push against the floor. This is necessary in order to move the upper body and your right shoulder against the wall. Because of the strain in your left leg it is impossible to move it away. If forced by someone else, you will fall towards the left, mainly because the **center of gravity** of your body is on the left side of your right foot, which would be the only support. In order not to fall, your center of gravity should be located exactly above the pivot point or support - in this case: your feet. After your left leg is knocked away, your body needs to move towards the right in order not to fall, but the wall prevents this.

Invitations to Science Inquiry – Supplement

FORCES   CENTER OF GRAVITY
                                          ADHESION

THE CENTER-SEEKING PAPER

Materials: 1. A small piece of writing paper (1/8 of letter size).   2. A wide, shallow tray.

Procedure:
1. Draw a small parallelogram (about 2x3 cm) with its diagonals on the little piece of paper.
2. Wet the parallelogram (not the whole paper) with water by letting droplets fall from your wetted index finger one at a time.
3. Fill the tray with water, lift the piece of paper carefully and let it float on the water surface.
4. Touch the tip of your index finger to the water on the parallelogram at a spot other than the crossing of the diagonals. What is the paper doing?

Questions:
1. Why does the water not spread to the other parts of the paper?
2. What force keeps the water on the paper together?
3. When touching the wetted part of the paper near one edge, which way does the paper shape move to?
4. At what point does the finger end up above the paper shape?
5. What makes the wet paper shape move?
6. What makes the paper stop moving eventually?

Explanation:
   The water drops on the paper are held together by *cohesive forces.* This is the reason why the water is not spreading to the other parts of the paper.
   When the finger tip is touched to one side of the paper shape, it is pulled to the other side by a larger cohesive force (see Sketch A). When touching the opposite side, the movement of the paper goes opposite too (see Sketch B).
   Whenever the paper stops moving, all *adhesive and cohesive forces are in balance.* It is also the center of gravity of the wetted shape. This can be used as a special way to find the center of gravity of an irregular shape.

FORCES   CENTER OF GRAVITY
                                                      SHOCK ABSORBERS

THE STANDING MATCHBOX

Materials: 1. One full or almost full wooden matchbox.
If used for class activity: 10-20 matchboxes needed.

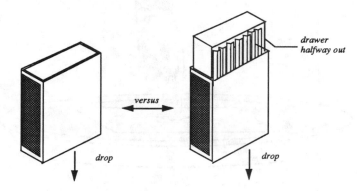

Procedure:
1. Distribute one matchbox per couple of students. Ask them to hold the box vertically above the table (about 15-20 cm above) and let go of the box. Then ask the question: "Can you drop it and leave it standing up?" or "Can you drop the box without it tipping over?" Give them time to try it out. Anticipated answer: "Impossible!"
2. Then you say: "Remember, nothing is impossible!"
   Take the box, open the drawer about half way out, hold the box vertically with the drawer up, and drop it from about the same height.
3. If you do not succeed in the first two or three tries, lower the height. Ask the students to try dropping the closed box ten times and counting the number of times it was left standing versus the half open box.

Questions:
1. Why does the closed matchbox almost invariably tip over when dropped?
2. What makes the half open box more stable?
3. What did we do to the center of gravity of the box by pulling the drawer half way out?
4. What happens to the center of gravity at the moment that the half open box hits the table surface?
5. How do people usually lessen the impact when jumping off a high ledge or loft?

Explanation:
As the matchbox is made out of cardboard, it is somewhat springy. Thus when dropped on a hard surface it is bound to bounce side ways, as it is almost impossible to let it land exactly flat on the smallest rectangular surface. Most of the times it lands on an edge or a corner of the box, and the box tips over. If the box was made out of clay, it would not bounce, it would actually give at the point of impact and stand up straight.
The half open box almost acts the same way as the clay cube. The outer sleeve acts like a cushion, as at the time of impact, the drawer, where most of the mass is, keeps sliding in further, thus breaking the fall. Its *center of gravity,* which is located in the center of the drawer, keeps moving in further at the point of impact. This is exactly how people break their fall when jumping off high places, by bending their knees just after or at the point of impact.

FORCES   CENTER OF GRAVITY
                                                    AIR RESISTANCE

STAND A DOLLAR BILL ON YOUR FINGER

Materials: 1. A crisp dollar bill, or any other strip of paper of the same size or somewhat larger.

Procedure:
1. Take the crisp dollar bill and stand it vertically on the table top. Show the audience that it is pretty difficult to keep it standing for any length of time.
2. Ask the students: "How long do you think can I keep it standing on my stretched-out finger?"
   Anticipated answer: "About 2-3 seconds".
3. Now bend/curve the bill somewhat length-wise and place it on your finger. Watch the tip of the bill and balance it (move your finger quickly to the left if the tip of the bill moves left, and to the right as soon as the tip of the bill moves right).
4. Let the students measure the time now of how long you could keep the bill standing on your index finger.

Questions:
1. How long can you balance tne dollar bill on your finger?
2. ls there a limit to the length of time you can balance it?
3. What keeps the bill from falllng over?
4. Would a longer or shorter strip of paper be easier to balance?
5. Would a wider or thinner (narrower) strip be easier to balance?

Explanation:
This balancing act can also be performed with any sheet of writing paper, as long as there is a slight bend in the paper to make it stand rather stable on your hand (or finger). The larger the area of the paper, the higher the *air resistance* when moving it from left to right or the other way. There is a cushion of air surrounding it, holding it up, as it were. The taller the strip of paper, the higher the center or gravity of the piece, and thus the easier the balancing. The wider the strip of paper, the larger the surface area, and thus also the easier to balance it, as the air friction (or cushion) will keep it vertical.

In juggling acts where the juggler balances a vertical pole on his head, the more mass is carried at the top of the pole, the easier it is to balance the whole system. This is because the top end will then be more *inert,* and it is easier to move the bottom end without moving the top end, and thus to move and place the support point (pivot) under the center of gravity of the system.(see **Another Balancing Act - page**).

Invitations to Science Inquiry – Supplement – Page 105

FORCES   CENTER OF GRAVITY
                                               SPINNING OBJECTS

KICK A STRAIGHT LINE

Materials: 1. A wood block (2x4" about 1 ft long) or any other shape.
  2. Four ballpoint pens.   3. A long piece of newsprint paper (or brown wrapping paper).

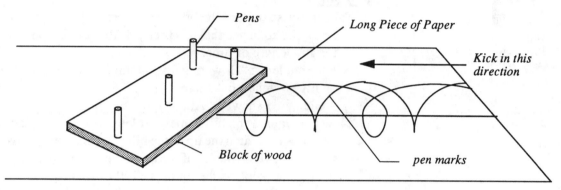

Procedure:
  1. Drill three holes in the wood block: two at one end and one at the opposite end. The size of the holes should be such that the pens can be inserted rather tightly in the holes.
  2. Insert three pens in the three holes, such that the point just sticks out on the other side of the board.
  3. Now find the center of gravity of the total system by finding the balance point on the tip of your finger. Mark that point.
  4. Drill the same size hole through this point and insert the final fourth pen (you can use a red one here) the same way (make sure that all four pens are evenly touching the table top).
  5. Place the wood block with the pen points on the long stretched out paper in the hall way.
  6. Kick the block of wood, such that you send it spinning down the hall over the paper (kick towards the end of the block). Observe the pen marks!

Questions:
  1. What kind of lines do you observe?
  2. Why did only one line stay straight?
  3. What kind of tracks would we get with only the first three pens inserted?
  4. Would there be any difference in tracks if we eliminated one of the first three pens? (by retracting it such that the point does not protrude enough to make a track, but leave the pen in the hole).
  5. What would happen to the straight line track if we took one of the first three pens completely out of the hole?

Explanation:
   This demonstration is another proof that *spinning objects spin around its center of gravity*. The fourth hole in the block of wood was made exactly in its center of gravity. By kicking the block against one end in the direction of the long piece of paper, it was spinning down the hall way around its *center of gravity*. Thus the fourth pen did not move side ways, and a straight line was obtained. Without this pen no straight line could be obtained.
   If one of the first three pens is completely removed, the fourth pen actually would not be located exactly at the system's center of gravity anymore, thus the straight line becomes also a wobbly line.
(see also **THE WOBBLING CIRCLES** - ISI-2nd Ed)

FORCES   CENTER OF GRAVITY
                                                                     SPINNING OBJECTS

STAND A RAW EGG ON ITS HEAD

Materials: 1. One raw chicken egg.   2. One hard-boiled chicken egg.
3. A piece of paper towel or handkerchief.

Procedure:
1. Place both eggs on the table and ask students: "How can I tell which one of the two eggs is hard boiled, without breaking the eggs?"
2. Spin the eggs one at a time with the same force: which egg spins much easier and faster?
3. Ask a student to stand the raw egg on its head on the table (with a table cloth). If the table has no table cloth, place the egg on a paper towel, napkin, or handkerchief. **It is almost impossible to do this!**
4. Now hold the raw egg firmly in your hand and shake it vigorously for about 30 seconds, then place it immediately on its head on the table (on the paper or cloth). Voila! It stands!

Questions:
1. Why did the hard-boiled egg spin faster than the raw egg?
2. How else can we tell which of the two eggs was the raw one?
3. Where is the center of gravity of a raw egg located?
4. What was achieved by shaking the raw egg vigorously?
5. What made it possible for the raw egg to stand on its head?
6. Would it be possible to stand the hard-boiled egg on its head?
7. Would you expect it to be harder or easier to stand the hard-boiled egg on its head, as compared to the raw one?

Explanation:
A chicken egg consists of a yoke, egg white, and the shell. The yoke is where most mass is concentrated and thus the ***center of gravity*** of the egg is close to the yoke or in the yoke. When a raw egg is spun, the yoke swings out (see Sketch) and swings back to the other side, thus slowing down the rotation. When the egg is hard-boiled, the mass inside the egg is solid, thus the ***center of gravity*** is stationary and the egg can spin fast around this point.
Shaking a raw egg vigorously will make the yoke more mobile or movable inside the egg-white, thus after shaking, the yoke can move down lower in the head of the egg. This lowers the ***center of gravity*** and the egg is therefore much more ***stable***. Try to stand the egg on its smaller end: it's possible too! - because the yoke can now move down to the smaller end too!

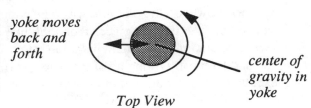

FORCES   ADHESIVE FORCES
                                                FRICTION

THE MAGIC STRIP OF NEWSPAPER

Materials: 1. A 3-4 cm wide strip of newspaper (about 30 cm long doubled). (unfolded strip: 60 cm long).
2. Rubber cement, talcum powder (or any kind of flour).

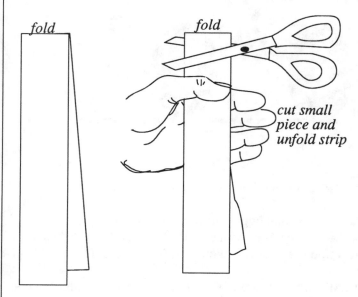

Procedure:
1. Unfold and lay the paper strip on a newspaper.
2. Cover one side of the strip with a layer (preferably 2 layers) of rubber cement and leave to dry.
3. Sprinkle some powder or flour over the dried rubber cement and wipe the excess off the strip (make sure all of the paper is covered with the powder).
4. Now you are ready for the demonstration: show the audience the paper strip.
5. Hold the strip close to the fold. Take a pair of scissors and cut about 2 cm off from the top, hold one end of the strip and hold it up: **it is one whole piece again!**
6. This cutting procedure may be repeated several times. Let students come up with a **hypothesis** of how this can happen.

Questions:
1. What makes the strip whole again?
2. After cutting, is the strip actually whole?
3. What is the function of the powder or flour?
4. What is the function of the rubber cement?
5. What other material can we use besides rubber cement?
6. What would happen if you cut the strip in an angle?
7. What hypotheses can you come up with to explain this event?

Explanation:
The function of the rubber cement is to make the two cut strips after cutting the long strip with the scissors, to **adhere** to each other. The powder or flour spread on top of the rubber cement makes the surface smooth and not sticky. The two rubber cement covered surfaces can then be placed on top of each other without sticking to each other. At the edge where the paper has been cut with the scissors, however, the pressure was high enough that just the edge and **only the edge** is sticking to each other. The two cut pieces then look like one whole piece again.

A variation to this demonstration is to cut both the top end and the bottom end, separate the top end and let it hang from one of the strip ends or leave both ends connected and show a loop. It is an excellent demonstration to make the students think and develop their imagination.

FORCES   FRICTION

THE INVISIBLE GLUE

Materials: 1. A plastic or glass bottle with a long tapered neck.   2. A piece of thin rope, a small cork.

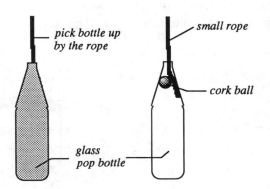

Procedure:
1. Preparation: Make a small ball (sphere) of the cork by cutting and filing until it just fits in the bottle neck (it should be a tiny bit larger so that you have to force the ball cork in).
2. Push the ball cork in the bottle and cover the bottle with paper or paint the whole bottle (so that it becomes opaque).
3. Now you are ready for the demonstration:
   Hold the rope in your hand and say to the audience: "I have some invisible glue in this small bottle" (an empty small bottle), "I'll dip the end of this rope in it and let it stick to this large bottle".
4. Do the dipping and let the rope slide inside the opaque bottle, turn the bottle upside down, pull slowly at the rope until you feel some resistance, then turn the bottle rightside up, keep the tension in the rope, and let the bottle hang from the rope. Let students make inferences/hypotheses.

Questions:
1. What hypotheses/inferences can you make to explain the event?
   After students know about the cork ball:
2. What is the function of the cork ball?
3. Why does it have to be somewhat larger than the bottle neck?
4. What other material can we use instead of cork?
5. How can we get a regular cork that went in, out of a wine bottle?
6. What is the principle or property made use of in this demonstration?

Explanation:
   The reason why the bottle can keep hanging from the loose rope is *friction*. When the bottle is turned upside down, the cork ball rolls in the bottle neck and pinches the rope against the side of the bottle. By pulling the rope, it pulls the cork in even tighter as the neck is tapered. The friction between the rope and the cork is greater than that between the glass and the cork, so the cork gets pulled a little farther in the neck, thus pinching the rope. The rope can be pulled out of the bottle with a sharp yank. In place of the cork we can also use a rubber ball. A marble of a slightly smaller diameter than that of the bottle neck may also be used, but the rope has to be quite a bit thicker.
   Making use of the same principle, we can take a cork out of an empty wine bottle with a cloth napkin (serviette). Get one of the corners of the napkin inside the bottle neck, let the cork roll over the napkin, slowly pull the napkin until the cork tightens in the neck, then pull hard on the napkin; the cork will pop out! (see **HOW TO GET THE CORK OUT II** - page 122)

Invitations to Science Inquiry – Supplement – Page 109

FORCES   WATER PRESSURE

THE CARDBOARD BOTTOM

Materials: 1. A glass or plastic open tube (open at both ends). (about 2-3 cm diam and 10-15 cm long).
2. A glass or other transparent container (4-500 ml beaker).
3. A stiff paper (or plastic) card, tape and thread, small beaker.

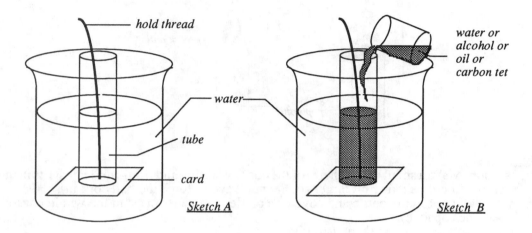

Sketch A   Sketch B

Procedure:
1. Tape the end of tne thread to the middle of the card, let the other end of the thread fall through the open tube, and hold the card by the thread against the bottom end of the tube (see Sketch A).
2. Lower the tube vertically in the water filled glass container, while holding the card against the bottom of the tube, until about half the length of the tube is immersed.
3. You can now let go of the thread (the card will be kept in place), and pour carefully and gradually some water in the tube from the small beaker (see Sketch B).
4. Continue to pour water in the tube until the moment that the card falls away from the tube.

Questions:
1. At what point does the card fall away from the tube?
2. What kept the card in place even after releasing the thread?
3. Why did the card not fall off even after pouring a little water inside?
4. Would the card fall off if we poured alcohol or oil in the tube instead of water? How about pouring carbon tetrachloride?
5. In order to pour the same amount of carbontet as the amount of water in the tube, what do we have to do with the tube (so the card would stay)?
6. Is it possible to pour carbontet up to the level of the water (without the card to fall off)?
7. What liquids can be poured in the tube up to the water level, without the card falling off?

Explanation:
Water pressure is exerted equally in all directions. Tne card falls off from the bottom of the tube only when the water column above it is flush (same height) with the water level outside the tube.

Lighter liquids (those of lower density) can stay up even higher above the water level, as it would take more of it to make the same weight of water. Heavier liquids will drop the card sooner. It is therefore impossible to fill the tube up to the water level with liquids of densities greater than 1. (see Sketch on right).

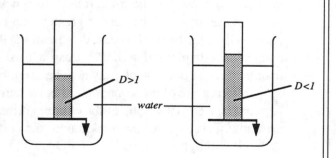

Invitations to Science Inquiry – Supplement – Page 110

FORCES   WATER PRESSURE

THE SQUIRTING WATER HOLES

Materials: 1. A large tin can (juice can) or milk carton.   2. A medium size nail, and bucket or sink.

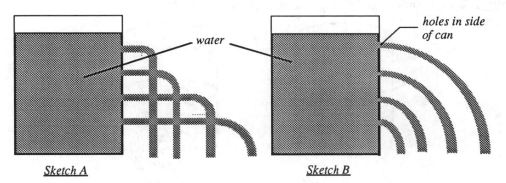

Sketch A   Sketch B

Procedure:
1. Punch 4 holes in the side of the tin can in a vertical line, about 3 cm apart from each other, the lowest one also about 3 cm from the bottom (make the holes the same size and shape).
2. Hold the can above the sink or bucket, have one of the students (X) cover the four holes with his/her fingers, and fill the can full with water.
3. Now ask the students: "Which hole will squirt water the farthest when student X takes away his/her fingers?", and let student X release the flow of water. Compare prediction and result.

Questions:
1. Which hole squirted the water out the farthest? In other words: which of the two sketches above is the correct one, A or B?
2. What factor determined the squirt distance from the can?
3. What are other variables that might influence the squirt distance?
4. How would different sizes of holes in the can compare?
5. Would the diameter of the can influence the squirt distance?
6. Would the height of the can or the height of the water level have more influence on the squirt distance?
7. How would water compare to alcohol or oil in the squirt distance?

Explanation:
The correct sketch above is Sketch A, showing that the top hole in the tin can would give the shortest squirt, and the lowest hole the farthest. This immediately shows that the height of the water level is influencing the squirt distance. The more viscous the liquid, the less the squirt distance, this means that oil would not squirt as fast as compared to water. A smaller hole compared to a large hole, with all other variables held constant (the same), will give a larger squirt distance.

Whether the liquid inside the can will squirt out at all, depends on the *liquid pressure* inside the can. This is caused only by the height of the liquid level above the point of puncture. This in turn is caused by the weight of the liquid or the gravity force. If there were no gravity working on a completely closed juice can, like in a satelite capsule, puncturing it at any place would not make the juice come out.

FORCES   WATER PRESSURE
                                                           VORTEX IN A LIQUID

THE OUTPOUR RACE

Materials: 1. Two identical gallon jugs (with narrow neck).   2. Sink or buckets.

Jug A   Jug B

Procedure:
1. Fill both jugs three quarters full of water. Place them next to each other and show the students that they have the same amount of water.
2. Ask two students to come forward to pour the water out as fast as they can into the sink or the buckets (student A already got instructions beforehand to swirl the water before pouring it out: *Hold your palm over the opening while holding the jug upside down, make a turning motion to swirl the water until it forms a vortex, then let go of your palm covering the opening*).
3. The rules are: The pourers may do anything to the jug before and during the pouring of the water (except break it).
4. Have two other students time the outpour with a stopwatch or by putting their hands up as soon as all the water has poured out.

Questions:
1. Which jug invariably wins? Jug A or Jug B?
2. What does the swirling do to the water level?
3. Why does it take so long for water to pour out of the jug?
4. What is trying hard to get into the jug while pouring the water out?
5. What is usually the cause of the vortex formation in the water surface?
6. What would you expect the pressure to be in the center of a vortex, higher or lower than the periphery?
7. In what places in nature can you find vortex formation?

Explanation:
By moving the jug in a clockwise or counter-clockwise rotation before pouring the water out, a ***vortex*** is formed, indicated by a funnel-like shape of the water surface. At that moment the center of the vortex has a much **lower pressure than the outer areas**, making it easier for the air to come into the jug. Thus the water will pour out faster. Keeping the water swirling while the water pours out will make it pour out steadily as the vortex is maintained.

Vortex formation is found in nature in fast flowing rivers, where the water has to go around a corner or around a large rock. In pools below a waterfall we can also find vortex formation, caused by strong currents of water. As the pressure in the center of the vortex is lower than its surrounding outer area, swimming near a waterfall can be quite dangerous. When caught in a vortex, the best thing to do to get out of it would be to go with the flow, which is downward and come up to the surface at another spot - *do not struggle to swim up to the surface as this will keep you under water!*.

FORCES   THE PENDULUM

HIT THE BOTTLE ON THE BACK SWING?

Materials: 1. A soft ball or base ball.   2. A thin string & hook or nail.
3. An empty coke bottle.

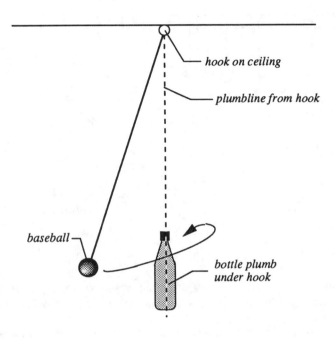

Procedure:
1. Tie the base ball to the string and hang it from the ceiling in such a way that it hangs about 2 cm higher than the bottle above the floor (do not make a knot in the string).
2. Let the ball hang very still and place the bottle plumb under the ball. Mark the position of the bottle on the floor with chalk.
3. Now lower the ball until it is about 10 cm above the floor, then put a permanent knot in the string (at the hook on the ceiling).
4. Bring the ball out towards you (about 2-3 m away from the bottle), swing the ball to the side of the bottle and try to hit the bottle with the ball on the back swing.

Questions:
1. Is it possible to hit the bottle on the back swing?
2. Why does the ball always swing back on the other side?
3. What is the energy of the swinging ball transformed into?
4. Will the ball eventually hit the bottle when it is allowed to swing more than once?
5. How should we swing the ball to obtain the least number of swings to hit the bottle?

Explanation:
The energy of the pendulum is transformed into heat at the point of attachment of the pendulum and in the air as friction. The amplitude or height of the swing is therefore almost the same for the first few swings.

When looking from the top of the pendulum and the bottle, we can decompose the movement of the pendulum along axes X and Y. As the swing stays almost constant along both axes, the ball can not hit the bottle on the back swing (for the first few swings). After a while the ball will fall back to the vertical position and eventually hit the bottle.

Invitations to Science Inquiry – Supplement – Page 113

FORCES   THE PENDULUM

HOW MANY SWINGS CAN YOU GET?

Materials: 1. Fifteen washers (for a class of 30).   2. Thin string or strong thread, and scissors.
3. Wooden meter stick, a dozen small hooks (to place in ruler).

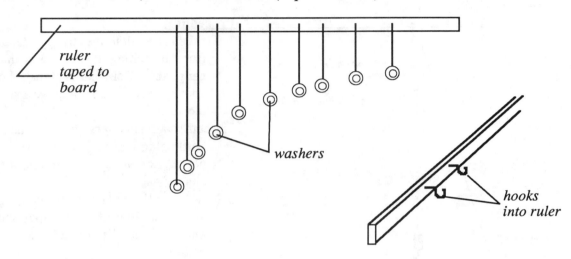

Procedure:
1. Place the small hooks into the wooden meter stick at the following spots: 39, 40, 42, 44, 47, 50, 54, 60, 66, 74, 94. To be safe, you may want to add a few more hooks about half or one cm away from each of the above points, then tape the ruler against the board (up high).
2. Tie a string to each of the 15 washers and make a knot at the washer and another knot the following lengths away: 60, 55, 50, 45, 40, 35, 30, 25, 20, 15, and 10 cm, and cut the string off 2 cm above this last knot, and make a small loop (to hang the pendulum on the hook).
3. Give each pair of students one pendulum. Let one student hold the pendulum with a steady hand, and let the other count the number of swings (whole periods) per minute (count the number of times it comes back to your hand in 30 seconds, then double it). Let them do this two or three times and take the average of the value.
4. Once they have the number, let them come to the board and hang their pendulum at the correct spot on the ruler (similar to plotting length of pendulum against frequency of swings).

Questions:
1. Which of the pendulums got the most number of swings?
2. What determines the number of swings per minute of a pendulum?
3. Would a different/heavier pendulum bob affect the frequency of swings?
4. Would a thinner or heavier string make a difference?
5. What are other variables that would affect the number of swings?
6. Which of the pendulums could be used as a timer?

Explanation:
This activity is especially effective to demonstrate that the number of swings per minute *(frequency)* in a pendulum depends solely on the length of the pendulum. It is independent of the mass of the bob. Variables like air friction, friction of the suspension construction, movement of the suspension point (in our case whether the student has a steady hand or not), may have the same influence on the slowing down of the pendulum and thus on the number of swings.

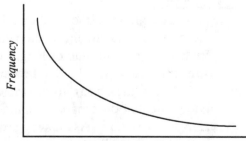

When plotting the number of swings per minute or frequency against the length of the pendulum, we will get a very similar parabolical curve as the one on the board.

FORCES
ELASTICITY
CONSERVATION OF MOMENTUM

IS THE BALL REPELLED?

Materials: 1. A strong rod magnet (or horseshoe magnet).
2. A small steel ball and a larger steel ball or two small identical steel balls.

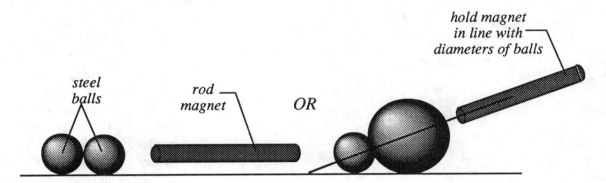

Procedure:
1. Show to the students that the two steel balls are both attracted by the strong magnet, and that both can easily hang on to the magnet.
2. Place the two balls on a smooth surface, next to and touching each other.
3. Approach the balls slowly with the magnet from a direction which is in line with the two diameters of the steel balls (if one ball is larger than the other, approach the larger ball with the magnet).
4. Ask: "What made the second ball move away from the first one?" A clue: if the magnet is not held in line with the extension of the two diameters, the second ball will not be moving away!

Questions:
1. What would the balls do when they are approached by the magnet in an angle to the two diameters?
2. What would the balls do when they are approached from the smaller one (in case of two different sized balls)?
3. What would you expect the balls to do when using an overly strong magnet? Would they behave the same way?
4. How close does the magnet have to be before the balls are attracted?
5. Why do the two balls have to be touching each other? What would happen if they were not?

Explanation:
The first hypothesis that an observer will form when seeing the second ball moving away from the first is, that it does this because it is repelled by the magnet somehow (by a change of polarity, which is completely erroneous!).
The actual explanation is that both balls are attracted by the magnet and at the time of impact the second ball is bounced back because of the ***momentum*** it has and the ***elasticity of steel metal***. It moves away out of the magnetic field and is thus not attracted any more by the magnet. This bounce will not occur when the magnet is not held in line with the two diameters of the balls. An overly strong magnet will hold both balls in its magnetic field and the bounce will also not occur.

FORCES   RIGIDITY

PIERCE A POTATO WITH A STRAW?

Materials: 1. A large raw potato.   2. Two regular drinking straws.

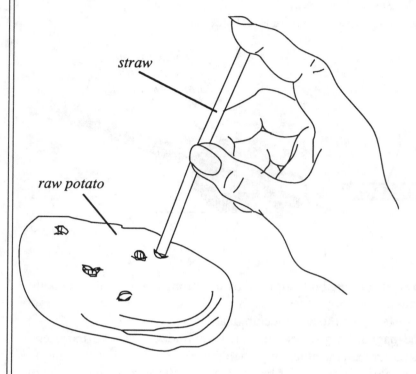

Procedure:
1. Hold one straw in your hand and push it against a hard surface, like f.i. the table surface and show the students that the straw is bending/folding.
2. Now hold the large potato in your left hand and the straw in your right (when you are right handed) with the index finger tightly pressed against the end opening of the straw (see Sketch on left)
3. Pierce the potato with one forceful stroke of the right hand. The straw goes right through the potato! This may be repeated.

Questions:
1. Why does the straw bend when pushed against a hard surface?
2. What is the purpose of holding the forefinger tightly pressed against the end of the straw?
3. What would happen if the forefinger was not held there?
4. What is inside the straw at the moment that it hit the potato?
5. Would the pressure inside the straw be higher or lower than the atmospheric pressure?
6. What are other examples where a higher pressure of air or water inside the object would make it more rigid and strong?

Explanation:
The purpose of placing the forefinger tightly against the end of the straw is to prevent the air inside the straw from escaping. At the moment that the straw hits the potato, the air inside the straw is compressed and the ***higher pressure makes the straw rigid and strong.*** The flexible straw then acts like a solid spear and penetrates the potato very easily. If the forefinger was not placed at the end of the straw, it would just bend when hitting the potato with it, just like it did when hitting the table top with it.

Other examples where a higher pressure of air makes the object rigid are: bicycle and car tires, inflated plastic or canvas beds and toys. Higher pressure of water inside the fibers and capillaries of plant tissue makes them more rigid and strong: fresh looking plants that are properly watered as compared to limp and drooping leaves because of water shortage.

FORCES   STRENGTH OF
                                                                        WEAVED MATERIAL

THE LOOSE KNIFE SUPPORTS

Materials: 1. Four regular drinking glasses. 2. Three table knives.

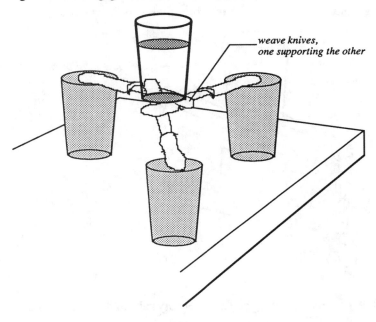

weave knives, one supporting the other

Procedure:
1. Place the three glasses in a triangle configuration on the table a little further than a knife long span away from each other.
2. Ask the audience: "How can I support the fourth glass in the middle of the other glasses at glass-height above the table with the three knives?"
3. Anticipated answer: "It's impossible!"
4. Give a clue: "Let one knife support the other!"
5. After everybody gives up: weave the knives as shown in the sketch above, and a strong support is created, which can hold the fourth glass quite easily. You may even want to fill this fourth glass full with water.

Questions:
1. How can the three knives make such a strong support?
2. How are the knife ends held up?
3. What does the weaving of the knives actually accomplish?
4. What other materials could we use instead of the knives?

Explanation:
   The end point of each of the knives is resting on the middle of the next knife, which the weaving structure actually accomplishes. By weaving the knives they act like a solid metal plate, which is certainly strong enough to hold up a considerable amount of weight. The structure is actually just as strong as the knife itself.
   Any other stiff material, like forks or spoons, or chop sticks may be used instead of the knives.

FORCES | STRENGTH OF CORRUGATED MATERIALS

THE DOLLAR BILL BRIDGE

Materials: 1. A dollar bill (any denomination or other currency).
2. Three drinking glasses (one empty, the others may be full).

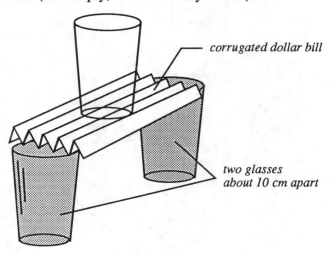

corrugated dollar bill

two glasses about 10 cm apart

Procedure:
1. Place two drinking glasses (that may be full or empty) about 4 cm less than the length of the dollar bill apart from each other.
2. Show people the third empty glass and the dollar bill. Ask: "Can you support the third glass at glass-height in between the two glasses?" Anticipated answer: "It s impossible!"
3. Say: "Nothing is impossible!" Proceed to fold the bill in half lengthwise then fold this in half again, and once more in half. Unfold the bill and fold it now in and the next crease out as in pleads of a skirt (in zig-zag pattern when seen from the side). Make sure that the folds are sharply creased (use your nail to press).
4. Spread the pleated bar somewhat out and place it over the two glasses, bridging the two, and carefully place the third glass in the center on top of the dollar bill. Voila! The dollar bill bridge!

Questions:
1. What made the dollar bill so much stronger?
2. Where do we see this similar structure in nature?
3. Can you find similar zig-zag structures (corrugated sheets) that strengthen different materials in our daily life?
4. Would more and smaller corrugations be stronger than fewer and larger corrugations in the dollar bill? Which is better?

Explanation:
Providing the dollar bill with the length-wise corrugations actually turned the bill into a strong beam across the two drinking glasses. There is an optimum number of corrugations which gives it an optimal strength. The larger the corrugations the less zig-zags we can make from the width of the dollar bill, and the more corrugatians the smaller they are and thus the weaker.

In nature we find similar corrugations in celery stalks and other plant tissues. In our daily life we encounter the zig-zag or corrugated structures in bridges, plastic roofing, cardboard boxes - cardboard consists of two layers of heavy paper with a corrugated layer in between.

FORCES   STRESSES IN PAPER

HOW LONG CAN YOU HOLD THE BURNING PAPER?

Materials: 1. A strip of paper (about 30 cm long, 4 cm wide). 2. A sharp edge of the table, and a match.

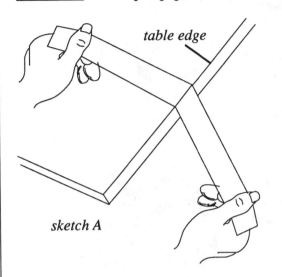

sketch A

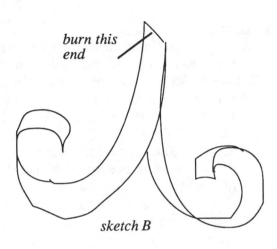

sketch B

Procedure:
1. Take the ends of the paper strip, one end in each hand, and rub it over a sharp edge of furniture (edge of table or chair). Do this 2 or 3 times (see Sketch A).
2. Notice that the paper strip curls up. Now fold the paper in the center in the opposite direction of the curl (see Sketch B), and hold the two ends together. (You may want to tape the ends together if you like to leave it lying on the table beforehand).
3. Now you are ready to ask a student: "If you hold the strip of paper (at the taped end) vertically, and I burn the top, how long do you think it will take before you have to let go of the paper?" Anticipated answer: "About half a minute".
4. Have a student hold the paper upright and burn the folded end. In about 3 seconds he/she will let go of the paper!

Questions:
1. What did the rubbing of the paper over the sharp edge do to it?
2. What made the student drop the paper so quickly?
3. How can we uncurl a curly strip of paper?
4. What did the burning of the folded end of the strip of paper do?
5. What would happen if we held the paper horizontally and burned it?
6. Would you hold the paper vertically downwards and start burning it from the bottom?

Explanation:
By rubbing the paper over the sharp edge, we caused the outside of the paper to stretch more, and thus a *tension* in the paper is created. When the paper is burned at the folded end, it separates and the two loose burning ends come curling downwards with lightning speed. Nobody would be holding on to a burning piece of paper if the flame suddenly moves towards the person. The immediate reaction of the person would be to let go of the burning paper.

Straightening out a curling strip of paper can be done by rubbing the outside of the curl over a sharp edge. Another way would be to apply heat to the piece of paper, by ironing for example.

FORCES   STACKING FORCES

THE STICKY KNIFE

Materials: 1. An empty jar (peanut butter/jam/coffee)
2. A medium-large, wide blade kitchen knife.
3. Raw rice grains (2-3 lbs)

Procedure:
1. Fill the jar almost completely full with the rice grains - mark the level of the grains on the side of the jar.
2. Take the knife by the handle and plunge the blade vertically into the rice grains. Show to the students that they are just loose grains, and that the knife can easily be lifted.
3. Plunge the knife several times into the jar of rice (4-5 times or sometimes even more) and slowly and steadily lift the knife vertically - the whole jar will lift with it! If the jar of rice is used for the first time, you may need to plunge the knife a few more times into the jar before it will be lifted by the knife.
4. Notice the level of the grains after the knife has been plunged several times into the rice grains. How did it change?

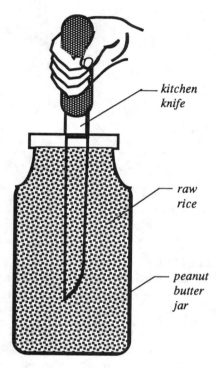

Questions:
1. How did the level of the rice grains in the jar change when we compare the first level with that after the knife has been plunged into it several times?
2. Did the resistance of plunging the knife into the grains change with the number of trials? How did it change? Why?
3. Why do we need to plunge the knife into the rice several times before it sticks?
4. What made the knife stick to the grains or the grains stick to the knife?
5. Will the same "sticking" occur in a jar with a wider opening than the bottom?
6. Where do the forces occur between the knife and the jar?
7. Can this phenomena occur by using other types of grain? Other knives?

Explanation:
   By plunging the knife for the first or second time into the rice grains, not much rearranging occurred between the grains, and that is why it is still easy to slide out the knife. After plunging it several more times into the grains, they get packed tighter. Because the knife is narrower and usually thinner, at one point the plunging of the knife pushes the packed grains towards the sides of the jar, and the grains push back towards the knife. The jar of grains act like one piece, just like a piece of wood, and the whole jar lifts with the knife. The force that the knife exert is against the sides and narrower part of the jar. This lifting cannot be done in jars with wider mouths.

Invitations to Science Inquiry – Supplement – Page 120

FORCES   NEWTON'S THIRD LAW

HOW CAN WE GET THE CORK OUT? (I)

Materials: 1. An unopened bottle of wine (corked).
            2. A cloth napkin or towel.   3. A solid wall (concrete or steel beam).

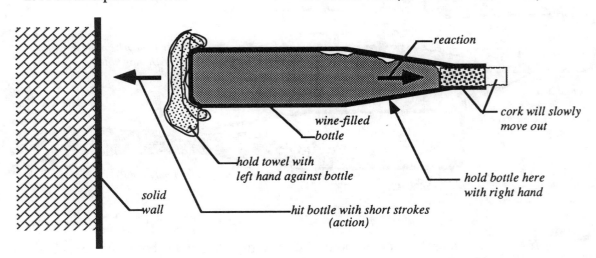

Procedure:
1. Wrap the towel around the bottom of the wine bottle, making sure that the towel provides a thick cushion on the bottom of the bottle.
2. Hold the bottle by the neck with your right hand, using your left hand to keep the towel against the bottom of the bottle, and hit the bottle against the solid wall with successive small thumps (see sketch above).
3. Observe the cork coming out of the bottle very gradually with every 3-4 hits!

Questions:
1. What function does the cloth napkin or towel have?
2. What can be used instead of the towel?
3. What made the cork move slowly out of the bottle?
4. Would this happen without wine in the bottle or with a half-filled bottle?
5. Will this procedure work better/easier with a wider or a narrower bottle (with the same mouth opening)?
6. What characteristic of the bottle is most critical?
7. Will a smaller (narrower) cork be easier or harder to take out of the bottle?

Explanation: *For every action* (force) against the wall with the bottom of the bottle, *there is an equal and opposite reaction* on the fluid (wine), which is transferred to the cork in the bottle mouth. The force exerted by the bottom against the wall was spread over the area of the bottom. The reaction, which is equal to this force, was spread over a much smaller area of the bottle opening (cork). This multiplied the pressure on the cork with a ratio of bottom area : neck opening area. The larger the bottom of the bottle or the smaller (narrower) the cork, the easier to open the cork using this method.

    The function of the cloth is to soften the impact of the glass bottle against the solid wall - in other words, to prevent shattering of the bottle. A wall that has a little of give to it, will not work for this demonstration (for example: regular gyprock walls). A concrete wall or steel beam/post is working best to get positive results!

Invitations to Science Inquiry – Supplement – Page 121

FORCES FRICTION

HOW CAN WE GET THE CORK OUT? (II)

Materials: 1. An empty wine bottle with its cork.   2. A cloth napkin.

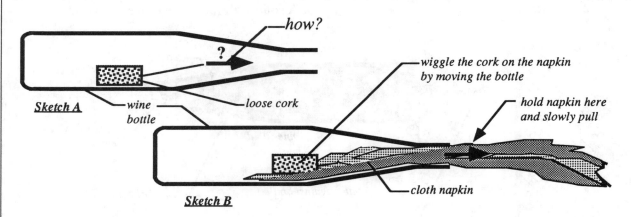

Procedure:
1. If the cork is loose outside the bottle, take it and push it in the bottle: completely inside the bottle.
2. Ask the students: "How can we take the cork out of the bottle, such that we can reuse the bottle and the cork?"
3. Take the napkin's corner, make a narrow chute out of it and push it as deep as possible inside the bottle.
4. Now wiggle the cork on top of the napkin inside the bottle.
5. Pull the napkin slowly out, making sure that the cork comes with it, until it gets tight (the cork is now pinched in the bottle neck with the napkin on one side of the cork).
6. Once the napkin is stuck between the cork and the bottle in the bottle neck, pull the napkin with a sharp yank the cork will shoot out!

Questions:
1. Why did the napkin not come out by itself?
2. What other ways would there be to get the cork out?
3. What other materials can we use instead of the cloth?
4. Is it easier to rub the cork against a glass surface or against the cloth surface?
5. Would an oily or wet cloth make it harder or easier to get the cork out? Why? Why not?

Explanation:
   This demonstration is particularly suited to get the students to use their imagination. Ask for suggestions as how to get the cork out of the bottle, without breaking the bottle or the cork. After students give up, take the cloth napkin and demonstrate. The cork comes toward the bottle neck when the cloth is slowly pulled out. The pulling of the cloth gets harder now, because the bottle neck is tapered. When the napkin is yanked out of the bottle, the cork comes with it, because the *friction* between cork and napkin is much greater than that between cork and glass (bottle). The rougher the cloth, the easier to get the cork to come with it. Oily or wet cloths will not be as effective to get the cork out.
   If it is a new bottle of wine, and we want to take the cork out without a cork screw, then we can do it by hitting the bottom of the bottle against a solid wall (see preceding event: "HOW CAN WE GET THE CORK OUT (I)")

FORCES   COMPRESSION
                                                                          WATER PRESSURE

THE WATER HAMMER

Materials: 1. A glass soda-pop bottle.
2. A rubber mallet (not essential).
3. A bucket.

Procedure:
1. Fill the bottle almost full with water - leave about one inch of space on the top.
2. Hold the bottle vertically by the neck close to the mouth with one hand and strike the opening with the mallet or palm of the other hand. Do this while holding the bottle over the bucket.
3. The bottom of the bottle will fall out! If this does not succeed, try again!

Questions:
1. What is the purpose of the air space above the water?
2. What does the water do when the bottle is hit?
3. Would the weight of the water affect this event?
4. Would this work with other liquids? Which?
5. Why does only the bottom of the bottle fall out?
6. Would this work if the bottle was only half or three quarters, or completely full?

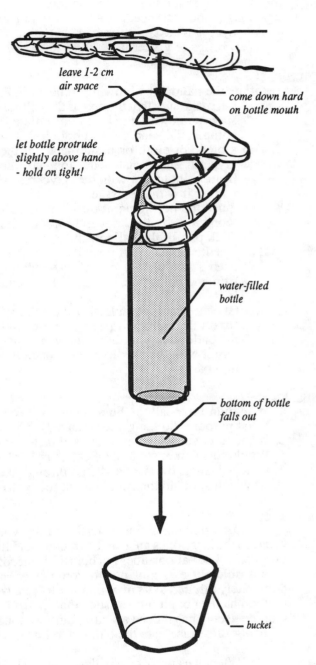

Explanation: The blow on top of the bottle moves the bottle suddenly downwards. Since there is a space above the water, the water moves up (relative to the bottle) - actually the water tended to stay at rest - *NEWTON'S LAW - FIRST PART : Objects at rest tend to stay at rest.* and this creates a vacuum immediately above the bottom of the bottle (on the inside) for a very short moment. Then suddenly the water comes back down with such a force that the water pressure pushes the bottom out.

If the bottle was totally full with water, there is no space for the water to move up in the bottle, no vacuum would be created, and no sudden pressure would be formed. If the bottle had less water, of course less weight would be pushing on the bottom and it will thus be harder to make the bottom fall out.

Invitations to Science Inquiry – Supplement – Page 123

FORCES   SHEARING

PUSH A NEEDLE THROUGH A BALLOON?

Materials: 1. A large round uninflated balloon.
2. A long (about 2 ft) needle (with or without an eye)
3. A few drops of glycerin.

Procedure:
1. Inflate the balloon by blowing some air in it, but make sure that the end of the balloon stays somewhat thick (do not inflate it completely!) and put a knot in it.
2. Dip a piece of cloth or tissue paper in the glycerin and moisten the long needle from point to eye with the glycerin by wiping it with the tissue.
3. Now push the needle point through the balloon close by the knot and through the thick part of the balloon end.
4. If the needle has an eye, thread it and pull the needle completely out of the balloon from the point side, but leave the thread in it.
5. Now take the needle completely off the thread (or off the balloon if it has no eye), hold the balloon in your left hand and puncture it with the needle point - touch the thin part - POOF!

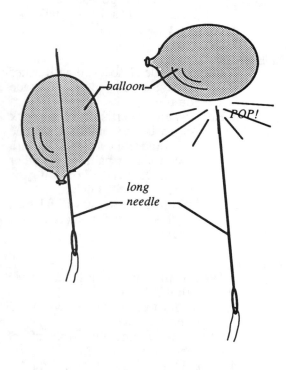

Questions:
1. Why didn't the balloon burst when the needle was pushed through the first time?
2. What purpose did the glycerin have?
3. What other substance may be used instead of glycerin?
4. Would this demonstration be easier or harder to do with small size balloons? Why?
5. At which spots did the needle go through the balloon?
6. At which spot did the needle point touch the balloon to make it burst?

Explanation:
       This demonstration is easier to do with large balloons as compared to smaller ones, because the large ones are usually thicker, and thus stronger. (Latex balloons are the best for this purpose). The balloon does not burst if the needle is pushed through the thicker parts, that is why it is absolutely essential not to completely inflate the balloon. Only when the balloon is completely inflated to its maximum will there be no thick parts left in it. The thicker parts of the balloon have enough rubber, such that *shearing* will not occur. When the time comes to show that the needle has indeed a sharp point, we touch the thin part of the balloon, and POOF! - the *shearing* force just tears the balloon to bits.

       When we approach a balloon with a sharp object, we usually cover our ears, anticipating a loud POP from the bursting of the balloon. Why is it that the bursting of a balloon is always accompanied with a loud popping sound? This happens because there is compressed air in the balloon and when this air is suddenly released - by the bursting of the balloon - a strong wave is created in the air, which reaches our ear drum and makes the ear drum vibrate.

FORCES   STRETCHABILITY

THE WATER-TIGHT ZIPLOCK BAG

Materials: 1. A small, plastic ziplock bag.   2. A couple of sharp pencils.

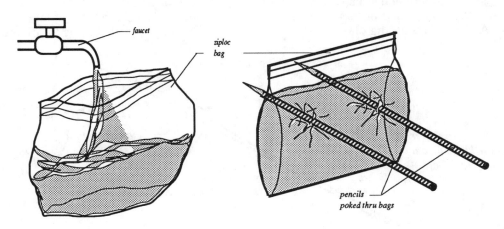

Procedure:
1. Fill the plastic ziplock bag almost full with water and close the zip tightly.
2. Show students that you can turn the bag upside down or lie it on the table without spilling the water out. Tell them that the bag is indeed so water-tight that we even can puncture it with a pencil without spilling the water.
3. Hold the ziplock bag in your left hand and push the first pencil somewhat on the side through the bag with one straight stroke, making sure that you do not move the pencil sideways!
4. Once the first pencil is completely through the bag, push the second pencil through with the same swift straight stroke (like throwing a spear at a fish).
   Practice this jabbing action with the pencil several times over the sink or a bucket!

Questions:
1. Why does the pencil have to be pushed through the bag with a straight stroke?
2. How can we stop the leak, once we made a hole in the bag with the pencil?
3. What other objects can we use to replace the pencil?
4. When the pencil is jabbed straight through the bag, why does the water not leak out?
5. Could we use a drinking straw to puncture the bag? When holding the straw with a finger at the end, would there be water in the straw after puncturing the bag of water?

Explanation:
When the plastic bag of water is punctured with the pencil with a straight stroke, no water will be spilled out of the bag. This is because plastic, which is polyethylene, is stretchable. This means that the hole made by the pencil is actually smaller than the diameter of the pencil. The plastic actually folds around the pencil and the water pressure presses the plastic against the pencil so that no water can leak out of the bag. If a small sideways force is applied to the pencil, the plastic tears, and a larger hole is made in the bag, resulting in a water leak. In order to stop a leak, once a hole is made in the bag, a larger diameter pencil or dowel (or any other cylindrical object) may be pushed into the hole.

When a straw is used to puncture the water-filled bag, it is essential to hold a finger against the end of the straw. When this straw is jabbed through the bag, at the moment that it hits the bag, air is trapped inside the straw, and it becomes rigid and acts like a solid rod. This means that after the straw went through the whole bag, no water is actually trapped inside the straw.

FORCES
FREE FALL
CENTER OF GRAVITY

THE MYSTERIOUSLY MOVING BALL

Materials: 1. A wooden stick (1"x 1/4" x3') about the size of a meter stick.
2. Two small plastic or paper cups.   3. Thumb tacks & Masking tape.
4. A blob of moulding clay.   5. A steel or plastic ball (3/4" diameter)

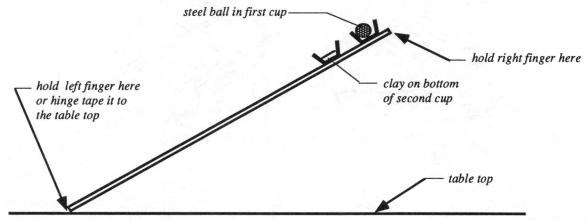

Procedure:
1. Attach the two cups at one end of the stick about 2" (5 cm) from each other with the thumb tacks and the masking tape. (see Sketch).
2. Tape the other end of the stick to the table top so that it hinges around the end (see Sketch).
3. Place a small blob of clay in the second cup (from the end) and spread it out on the bottom.
4. Place the ball in the end cup and lift the stick at the cup end about 2 ft (50 cm) (see Sketch).
5. Remove your finger that held the stick up with a sudden movement away from the stick end, in other words: let the stick end suddenly fall!
6. Observe the ball: it moved into the second cup! (If it does not fall in the second cup, vary the intitial height of the stick and practice letting go of the stick end)

Questions:
1. Did the ball experience free fall?
2. What point of the stick actually experienced free fall?
3. Was the very end of the stick falling slower or faster then a free fall?
4. How could the ball actually jump out of the first cup?
5. How far can the second cup be placed and yet still catch the falling ball?
6. What would be the optimum height of the stick end, so that the ball would move easily?
7. How can we test which point of the stick is actually experiencing free fall?
8. Will the ball always jump out of the cup no matter where the cup is placed on the stick?

Explanation: The *center of gravity* of the lath with the cups attatched is about 2/3rd away from the hinged end. This point falls at the same rate as the steel ball.

As point A (center of gravity) is not as high above the table as point B (h2>h1) point B has to fall faster than a regular *free fall*. Thus, the right-hand end of the lath indeed falls faster than the steel ball, and as the ball falls in a straight path, it ends up in the second cup.

This means then that when a hinged lath or stick *falls freely*, the end of the stick which is not higned, always falls faster than a free fall (assuming that the hinged end is not moving).

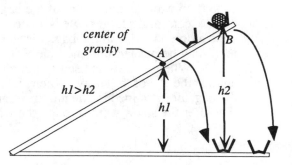

Invitations to Science Inquiry – Supplement – Page 126

FORCES   PERPENDICULAR FORCES

CRUSH THE CAN BY STANDING ON IT?

Materials: 1. An empty soda pop can.   2. A thick rubber band.

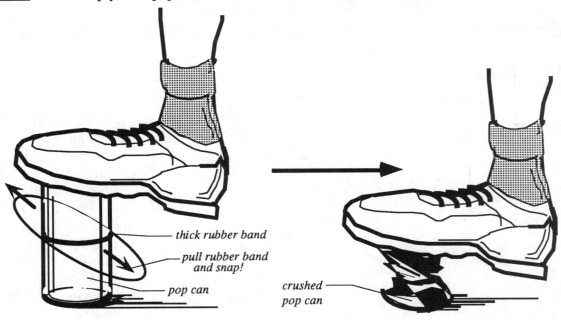

- thick rubber band
- pull rubber band and snap!
- pop can
- crushed pop can

Procedure:
1. Place the rubber band around the pop can in the center of the height.
2. Put the can on the floor near a table or chair (to hold on to), and have a student of about 100 lbs stand on the can with one foot, while holding on to the table or chair.
3. Now stand in front of the student face to face, squat down, pull the rubber band on both sides of the pop can and snap both sides at the same time! **CRUSH!**

Questions:
1. What was holding up the student's weight?
2. What would happen if the weight of the student standing on the can is too heavy?
3. What is the maximum weight that the pop can is able to hold before crushing? How can we find out?
4. What would happen if only one side of the rubber band was snapped?
5. What else could we do instead of using a rubber band?
6. In which direction does the rubber band exerts its force?
7. Can you name some examples where this particular principle is applied in daily life?

Explanation:
   Depending on the soda pop can, the maximum weight it can support is between 100 and 150 lbs (or between 60 and 90 kg). When we step on this pop can, our weight or the force is vertically downward. When the rubber band is snapped on both sides, it exerts a force perpen-dicular to the weight, and suddenly the can is not able to hold the weight, and the can crushes.
   Instead of the rubber band, a ruler can be used. Just a quick hit on the side of the can will make it crushing down. It is better to use two rulers and hit the can on both sides. If we use only one ruler or let the rubber band snap only on one side, the can may get crushed on one side moreso than on the other, and the student may fall to one side.
   We see this principle applied in daily life, when we see old tall buildings being demolished. The dynamite is placed near the bottom of the building so that the weight of the building itself will destroy it. It is very important to let the blasts go off exactly at the same time at all sides of the building, otherwise it may come down sideways and destroy other surrounding buildings.

Invitations to Science Inquiry – Supplement

FORCES   ROTATIONAL FORCES
                                                        TORQUES

THE CONFUSED TWIRLING PAPER

Materials: 1. A strip of paper (about 15 cm long, 5 cm wide).   2. A pair of small scissors.

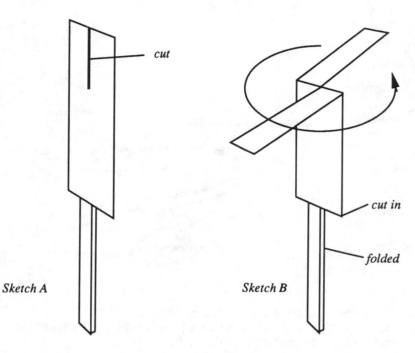

Sketch A   Sketch B

Procedure:
1. Take the paper strip and cut about 5 cm into the center of the strip and fold the two parts to opposite sides of the strip (see Sketch A).
2. At about 5 cm from the other end of the strip, cut cross wise one third into the width of the strip and fold the two parts on top of each other, making the bottom part thicker (see Sketch A).
3. Take the paper piece in the middle and drop it to the floor from standing height. Which way does the paper piece rotate? (See Sketch B)
4. Now fold the two "wings" to the opposite side. Drop the paper piece again from standing height to the floor. Which direction does it rotate?

Questions:
1. What makes the paper piece rotate when dropped to the floor?
2. What are the forces working on the paper piece?
3. What direction did it rotate the first time? And the second time?
4. What made the rotation change direction?
5. Can you find examples in nature making use of this principle?

Explanation:
The forces working on the paper piece when dropped to the floor are caused by air friction. When looking at the piece from above (obliquely, like in Sketch B above), and the right wing is folded away from you and the left towards you, dropping it will make it rotate clockwise.

When the "wings" are folded in the opposite way, the paper piece rotates counter clockwise. When we look at the forces working on the paper we note a torque working in both cases (see Sketch on right).

Many seeds of different fruits in nature are equipped with "wings" which will promote the propagation of the seeds when released from the ripe dry fruit.

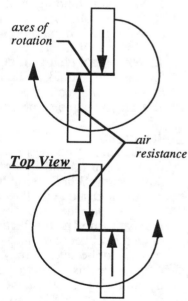

Top View

FORCES ROTATIONAL VIBRATIONS
DIRECTIONAL OSCILLATIONS

THE YIP-YIP STICK

Materials: 1. Two wooden dowels (one somewhat thinner than the other).
or two small pine sticks (about 1 x 1 x 15 cm). 2. A coffee stirrer and a small nail.

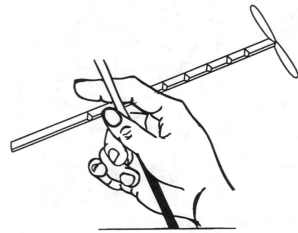

Procedure:
1. Make small notches on one edge of the pine stick about 1 cm apart from each other. The notches should be about 3 mm deep and about 2 mm wide.
2. Cut a 6 cm piece off from the coffee stirrer and make the same rounding edge on the cut end.
3. Nail the middle of this piece of stirrer to the top end of the pine stick, and wiggle it so that it will turn loosely around the nail (the nail should be hammered in only half way).
4. Hold the smaller stick in your right hand (if you are right handed) with your index finger pointing out (see Sketch) and hold the notched stick in your left hand.
5. Rub the notched stick with the smaller stick over the notches and let your thumb slide on the right side of the notched stick (stirrer will rotate counter-clockwise as seen from the performer's viewpoint).
6. Tell the audience that the stirrer (rotor) will turn the opposite way when you say "Yip-yip". Say "Yip-yip" and keep rubbing the notched stick but this time let your index finger slide against the notched stick. Observe the rotor rotate in a clockwise direction!

Questions:
1. What made the rotor turn?
2. What caused it to turn one way and what caused it to turn in reverse?
3. Did the saying of "Yip-yip" have anything to do with turning it in reverse?
4. Does the distance between the notches have an influence on the rotation?
5. Does the speed of the rubbing have anything to do with it?
6. Would rubbing the notched stick on one side, then turning it to the other side, have anything to do with it?
7. What would happen if you held the smaller stick another way?

Explanation:
The rubbing of the smaller stick against the notched stick with the rotor on, produces vibrations. When the thumb is held against the notched stick while rubbing, the oscillations take the form of an elipse causing the stick to vibrate in a counter-clockwise direction, and thus the rotor rotates that way too.

When the index finger is held against the notched stick, the oscillations are slanted in a clockwise direction. This changing of rotational direction can also be achieved by turning the notched stick slightly to the left or to the right, in other words by rubbing the notches on the left side as compared to rubbing it on the right side.

Invitations to Science Inquiry – Supplement – Page 129

FORCES — PRESSURE VS FORCE

CRACK A NUT

Materials: 1. Two wallnuts.

Procedure:
1. Hold one wallnut in your hand and try to crack it open without using any utensils. You will find that it is almost impossible to do!
2. Hold both wallnuts in the same hand (see Sketch) and use the same force to crack the nuts open: one of the nuts cracks!

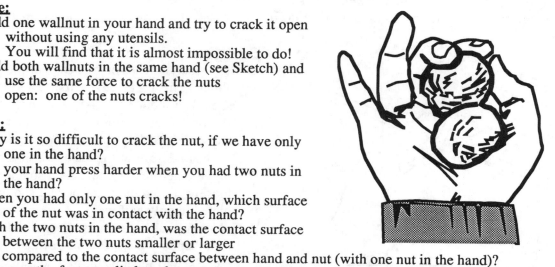

Questions:
1. Why is it so difficult to crack the nut, if we have only one in the hand?
2. Did your hand press harder when you had two nuts in the hand?
3. When you had only one nut in the hand, which surface of the nut was in contact with the hand?
4. With the two nuts in the hand, was the contact surface between the two nuts smaller or larger compared to the contact surface between hand and nut (with one nut in the hand)?
5. Why was the force applied on the nut not great enough to crack it with one nut in the hand?
6. Where is this principle of spreading the force over a larger surface area to reduce it being used in other situations in daily life?

Explanation: When only one nut was held in the hand, the force of the closing hand was spread over almost all of the nut surface (see Sketch Nut A). With two nuts in the hand, the contact surface between the two nuts is very small.

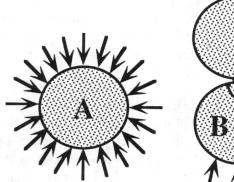

The force that was first spread over the upper half of the nut is now concentrated on one spot (see Sketch Nut B). Although the hand was actually exerting the same force with the one nut or with two nuts in the hand, the *pressure = force per square inch of surface area* is much greater on nut B (with two nuts), therefore it cracks open!

This same principle of reducing the force by spreading it over a larger surface is applied when helping someone who has fallen through thin ice in the winter. In order not to fall through the thin ice and yet come close to the drowning person, one has to lie down on the ice, spread arms and legs, and pull the person out of the freezing water.

Another example in daily life is in carpentry, when wanting to move a nice finished piece of wood a fraction of an inch with a hammer and we do not want to mar its surface, we take a piece of wood that has a larger surface than the hammer head, place it on the finished surface and hit the small piece of wood. This way no damage is done to the nice finished surface, because the hammer blows are not concentrated in one small spot, but spread over a much larger area.

You might have seen thr trick where a person lies on a bed of nails, and a block of concrete placed on his body is smashed in half with a mall without harming the person. This is the same principle applied, where the person's weight is distributed over the whole body surface (the nail points are about an inch apart from each other), and thus the force of the hammer is spread over the large concrete block surface on one side and the body surface on the other side.

SPACE SCIENCE   NEWTON'S FIRST LAW

ANOTHER BALANCING ACT

Materials: 1. A long, thin wooden dowel (4-6 ft)   3. String or tape.
2. A heavy weight (kg mass or rock)

Procedure:
1. Tie or tape the heavy weight (in the form of weight mass or a heavy rock) to one end of the wooden dowel.
2. Place the other end of the dowel (without the weight) on your index finger, and balance the dowel vertically by moving your finger sideways, forward or backward as needed.
3. Now let a student try to balance the dowel by placing the weighted end on his index finger!
4. After failing to do so, let him turn the dowel upside down and balance it now! Why is it so much easier to hold the dowel vertical on the finger?

Questions:
1. What does Newton's Law say about objects at rest?
2. Would a hievier or lighter object have more innertia?
3. In order for the dowel to stay vertical, where does the center of gravity have to be in relation to the pivot point?
4. Where is the center of gravity of the system located? - When the weighted end is on top? When the weighted end is below?
5. What point do we move when we move our finger on which the dowel rests?
6. Can you find some examples in daily life where this principle is applied?

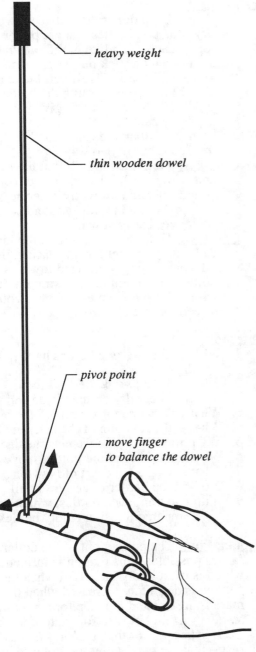

Explanation: The longer the dowel or the heavier theweight sitting on the high point of the dowel, the higher the *center of gravity* of the system. The higher the *center of gravity* of the system, the easier it is to shift the *pivot point* under the center of gravity. This means that it is easier to keep the dowel vertical on the index finger. A second reason for the ease to keep the dowel vertical with a heavy weight on top of it, is that heavier objects have greater innertia - *Newton's Fist Law - First Part says: Objects at rest tend to stay at rest..* This means that it is harder to move heavier objects, and thus is it easier to move the pivot point (this is the point where the dowel rests on the finger) under the center of gravity.

Examples of this principle in daily life we see constantly in circuses, where people balance different loads on top of a long pole. We also see this principle applied when acrobats ride unicycles: the higher the unicycle, the easier it is to ride it. People on stilts can balance themselves easier when the stilts are higher compared to using the shorter ones.

SPACE SCIENCE   NEWTON'S FIRST LAW

PULL THE DOLLAR BILL OUT?!

Materials: 1. Two identical, empty soda pop bottles.   2. A crisp dollar bill.

Procedure:
1. Place a crisp dollar bill between two dry soda pop bottles that are put on top of each other like in the sketch (let about an inch stick out on one side).
2. Ask the students: "How can I take the dollar bill without letting the bottle fall?" Ask the students to give you suggestions.
3. Then show that you can actually take the dollar out in two ways:

Option A: Hold the dollar bill at the long end with your left hand, keep the bill straight out and hit with a ruler or with your right-hand middle and index finger vertically down.

Option B: Hold the dollar bill at the long end with your right hand, slacken the bill (by moving your hand towards the bottles), then pull the bill straight out with a sharp movement in a horizontal direction. (see sketch)

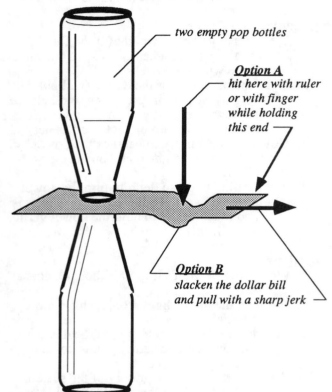

two empty pop bottles

Option A
hit here with ruler or with finger while holding this end

Option B
slacken the dollar bill and pull with a sharp jerk

Questions:
1. Why do the two pop bottles have to be dry?
2. Which of the two bottles has to be definitely dry on the inside and the outside?
3. What will happen if we pulled the bill using wet bottles?
4. What is the function of the ruler in Option A?
5. In Option B, why do we need to slacken the bill first before pulling?
6. Why did neither of the bottles move with the dollar bill?
7. What does the first part of Newton's First Law state?
8. Would it be harder or easier to pull the bill out with two filled (unopened) bottles?
9. With two unopened bottles, which two factors are increased when pulling the bill out, as compared to using empty bottles?

Explanation: The pulling out of the dollar bill from in between the two bottles has to be done as fast as possible. This motion is influenced by the *friction* between the paper and the glass rims, which is most likely increased when the bottles are wet - especially the top one. This factor of *friction* is definitely increased when the bottles are unopened (with their caps on), because their mass is greater and thus the force pressing on the dollar bill is also greater. The *innertia* or *tendency to stay at rest* for each of the bottles, however, is also much greater.

It is because of this *tendency to stay at rest* - **Newton's Law: Objects at rest tend to stay at rest** - that the two bottles are not moving with the dollar bill when it is moved, provided it is moved out very quickly. The function of the ruler in Option A is to snatch the bill out much quicker than the fingers can. A regular fast pull of the bill is most likely not fast enough and both bottles will move with it. In Option B there is more room (distance) for the hand to move, which prevents the bill to undergo the initial velocity of the hand movement.

Applications of this principle we encounter in tearing a piece of toilet paper off the roll with one hand and when magicians pull table cloths from under dishes and silverware.

Invitations to Science Inquiry – Supplement – Page 132

SPACE SCIENCE

INERTIA
NEWTON'S FIRST LAW

GET THE EGG IN THE GLASS

Materials:
1. One raw egg.
2. One egg holder.
3. One aluminum pie plate.
4. A large drink glass.
5. A household broom.

Procedure:
1. Fill the drink glass three quarters full with water.
2. Put the empty pie plate on the glass and center the egg holder and raw egg on the pie plate directly over the glass.
3. Place the glass with everything else on top, close to the edge of the table such that the rim of the pie plate hangs over the table edge (see Sketch).
4. Ask the students: "How can I get the egg in the glass with the broom without breaking it?" Anticipated answer: "Have no idea!"
5. Hold the broom directly in front of the set up, push down on it so the sweep part of the broom bends. Place one foot on the broom while holding back the wooden handle, then suddenly release the pole and flick it hard against the pie plate (see Sketch).

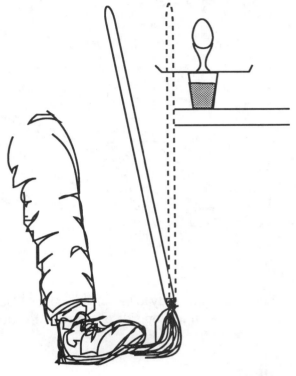

Questions:
1. What made the egg fall exactly in the glass?
2. What was the function of the water in the glass?
3. What would happen if we had a plate without a rim on the glass?
4. Why did the pie plate have to protrude over the table edge?
5. Would any other object in the egg holder end up in the glass?

Explanation:
The broom stick hit the edge of the pie plate without hitting the glass or the egg, because its motion was stopped by the table edge.

This sudden force moved the pie plate out from under the egg carrying with it the egg holder, as this was caught by the plate's rim (see Sketch). The egg was *at rest and tended to stay at rest (first part of Newton's First Law)*. The water was needed to catch the raw egg and prevent it from breaking. Any other object on the egg holder would have fallen in the glass, provided it has a heavy enough mass (and thus have enough inertia).

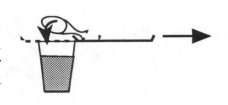

Invitations to Science Inquiry – Supplement

SPACE SCIENCE   NEWTON'S THIRD LAW

THE STRAW ROCKET

Materials: 1. A plastic flexible bottle.   2. Two drinking straws of two different sizes.
3. Some molding clay and construction paper.

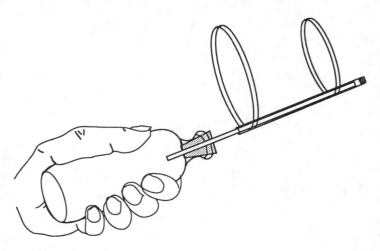

Procedure:
1. Prepare the bottle starter by putting a roll of molding clay around one end of the smaller straw and fitting it into the mouth of the plastic flexible bottle - test for leaks by plugging the other end of the straw with a finger, then squeezing the bottle: this should feel hard to squeeze when there are no leaks.
2. Prepare the rocket by taping two paper loops to the larger straw. One smaller loop to the front end and a larger loop to the back end. Stuff the front end of the straw with a blob of clay.
3. You are now ready to launch the rocket. Place the larger straw over the smaller one, hold the paper loops on the top side of the rocket with another straw, and squeeze the bottle with a sudden motion.

Questions:
1. What made the rocket move forward?
2. Why does the smaller straw have to fit tightly in the bottle?
3. What is the function of the blob of clay in the larger straw?
4. What is the function of the paper loops?
5. What initial action gave the rocket its energy to shoot forward?

Explanation:

By squeezing the plastic bottle, air is blown through the smaller straw into the larger straw. As this latter one is plugged at the front end, a higher pressure is built up in the larger straw and this makes the rocket shoot forward. As Newton s Law states: *For every action there is an equal and opposite reaction.* The action here is air shooting out of the rear of the rocket (larger straw), and the reaction is that the rocket moves forward. The function of the paper loops is to keep the straw floating in a horizontal direction.

Toys that shoot plastic balls by squeezing a plastic gun, and air guns (Beebee guns) are applications of this basic principle of *action is reaction.*

SPACE SCIENCE CENTRIPETAL FORCE
GRAVITATION

THE GRAVITY MACHINE (I)

Materials: 1. An old variable speed turntable. 2. Two identical empty jam jars (with lid).
3. A wooden stick (double the width and length of a 30 cm ruler).
4. Two medium size candles. 5. Epoxy glue or some other strong cement.

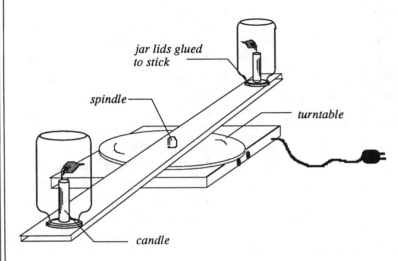

Procedure:
1. Make a small hole in the center of the wooden stick, such that it will fit tightly over the spindle of the turntable.
2. Cement the two jar lids upside down on each end of the wooden stick with epoxy glue or other strong cement.
3. Attach each candle to the center of each of the jar lids (with candle wax).
4. Light the candles and show what the flames are doing when the turntable is switched on (use the low speed for this).
5. Ask students to predict what the candle flames will do when the glass jars are screwed on just before the table is turned on.
6. Screw the glass jars over the burning candles into the lids and immediately switch on the turntable (using higher speeds). Observe the flames!

Questions:
1. What made the flames point towards the center of the turntable?
2. Which way does a candle flame point in still air? Why?
3. What does the spinning of the jars do to the air molecules in the jar?
4. What part of the jars would contain denser air?
5. In which direction did the spinning create a gravitational force?

Explanation:
The spinning of the jars on the turntable created a centripetal force (center seeking). The air molecules in the jars are put in motion and tend to stay moving in a straight line. As the jars move in a circular motion, the air molecules tend to move toward the outside (farthest from the spindle), thus making the outside part of the jars more dense. Since *candle flames always point towards less dense areas of air* - in still air they point upwards as the heat makes the air above the flame less dense - in the jars they point towards the inside part or towards the center of the turntable.

In space, satellite stations' gravity will most likely be created in the same fashion. Other applications are found in amusement parks where we find rotating drums and motorcycles riding against the walls of a drumlike pit.

SPACE SCIENCE   CENTRIPETAL FORCE
                                                         GRAVITATION

THE GRAVITY MACHINE (II)

Materials: 1. Two identical empty jam jars (with lid).   2. Two ping-pong balls, thread, and tape.
3. A sturdy wooden board.   4. A turntable or "lazy susan".

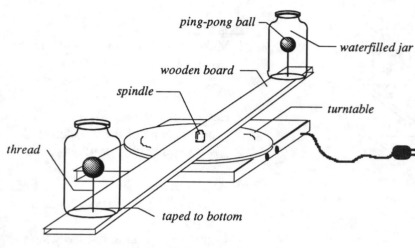

Procedure:
1. Drill a small hole in the center of the wooden board, such that it will fit tightly over the spindle of the turntable; and fit the board on it.
2. Tape the two jars right side up against the ends of the board (see Sketch).
3. Attach a short piece (about 10 cm) of thread to each of the ping-pong balls, and attach the other end of the thread against the bottom of each of the jars (make sure that when the thread is held up tight, the ball will only come up to about three quarters of the jar's height).
4. Fill the jars full with water (the ping-pong balls should be totally submerged under water and hanging up from the thread), and close tightly.
5. Before rotating, ask: "Which way will the ping-pong balls move, when the table rotates?" Anticipated answer: "Outward, or against the direction of the rotation". Now switch the turntable on, and observe! (or turn the lazy susan).

Questions:
1. Which way did the ping-pong balls move towards during rotation?
2. Why did the ping-pong balls hang upwards in the water?
3. What would you expect the ping-pong balls to do if they were hung from the lids in air-filled jars?
4. What would golf balls do, when hung by threads from the lids in waterfilled jars? In air-filled jars?
5. Which way would a helium-filled balloon (tied to a thread) in an accelerating bus move to? Forwards or backwards or stay stationary?

Explanation:
Just like the candle flames in the former Event: **THE GRAVITY MACHINE I**, the ping-pong balls are lighter than water, and *tend to float towards areas of lesser density.* As soon as the water-filled jars are being rotated, the water molecules are flung outward, making the outside of the jar denser and the inside (toward the center of the turntable) less dense. This is the cause of the balls to be flung inwards.

If the ping-pong balls were suspended from the lid in air-filled jars, the balls would fling outwards, as they are heavier than air. So will golf balls in water or air behave. But helium-filled balloons in a suddenly moving bus would move forward when the bus accelerates, and move backwards when the bus suddenly stops.

SPACE SCIENCE   CONSERVATION OF MOMENTUM

HOW HIGH WILL THE BALL BOUNCE?

Materials: 1. One golf ball.   2. One ping-pong ball.

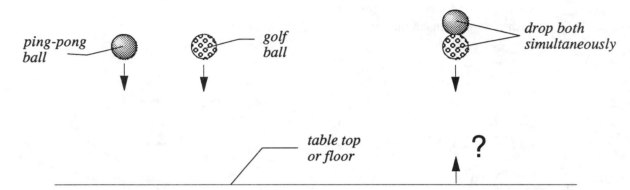

Procedure:
1. Hold the golf ball about waist high above a smooth hard floor or about half a meter above the table, and let it fall. Let students observe the height of the first bounce.
2. Do the same with the ping-pong ball and let the students again notice the height of the first bounce.
3. Now place the ping-pong ball on top of the golf ball and ask the students to predict how high the ping-pong ball will bounce up.
4. Drop the two balls simultaneously on the hard surface and observe!
(sometimes the ping-pong ball will shoot off in an angle; repeat until both balls fall vertically on top of each other).
5. Invert the order of the balls and ask: "What do you expect the balls to do now?" Drop the balls and observe!

Questions:
1. What made the ping-pong ball bounce up so high?
2. Was the height of the bounce the sum of the heights of the separate balls bouncing? Was it higher than the sum?
3. Where did the ping-pong ball get its energy from, to bounce so high?
4. Why did the balls hardly bounce up when the golf ball was on top?
5. What would happen if we used a tennis ball instead of the golf ball?
6. How would the bounce of two ping-pong balls compare to our first bounce?

Explanation:
The height of the bounce of each individual ball is about three quarters of the original height. When the ping-pong ball falls together with the golf ball (vertically underneath it), one would expect the bounce to be the sum of each ball individually making it twice as high, but it bounces up much higher. This is caused by the much larger mass of the golf ball as compared to the ping-pong ball. As the *momentum* (of the golf ball is *conserved*, it is actually imparted on the ping-pong ball : ($M_1 v_1 = M_2 V_2$). Because the mass of the ping-pong ball is much smaller, it has to have a much larger velocity (V_2) in order for the product of mass and velocity to stay the same.

When the two balls are inverted, there is only a small momentum imparted on the golf ball after the bounce, which is not large enough to bounce the golf ball back, thus both balls fall dead on the surface.

SPACE SCIENCE

CONSERVATION OF MOMENTUM
RECOIL

THE TEST TUBE CANNON

Materials: 1. A thick walled test tube + cork or rubber stopper.
2. Tall stand, thin wire, Bunsen burner.   3. Protective goggles.

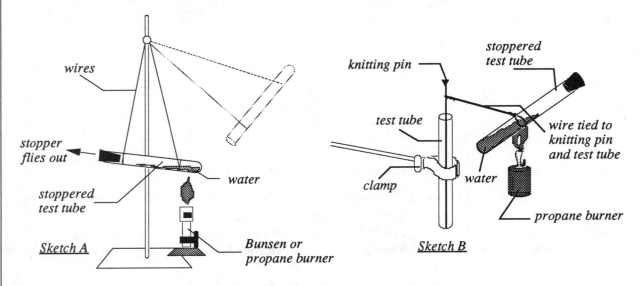

Procedure:
1. Tie two wires to the test tube and hang it from the tall stand as shown in Sketch A (somewhat slanted with opening up).
2. Pour a few ml of water in the tube and stopper with the cork or solid rubber stopper. **CAUTION: DO NOT INSERT STOPPER TOO TIGHTLY!!**
3. Put your safety goggles on and start heating the tube. Make sure that the mouth of the tube is facing away from people and breakable material.
4. Keep heating to boil the water - hold the burner with an outstretched arm and keep your face as far as possible from the tube, and wait for the POP! Observe the movement of the test tube!
5. An alternative way for suspending the test tube is shown in Sketch B.

Questions:
1. What made the test tube move at the time of the "POP"?
2. How does the use of a rubber stopper compare to a cork?
3. Will the use of more or less water influence the amount of recoil?
4. How would a longer or wider test tube influence the recoil?
5. What is the danger in inserting the stopper too tightly?
6. Will the tightness of the stopper influence the amount of recoil?

Explanation:
      By boiling the water in the test tube, vapor is being created, and thus pressure is built up, as the test tube is stoppered tightly. The vapor molecules move faster and faster as energy is supplied to the steam by the heat, and at the moment that the pressure is high enough to overcome the friction of the stopper against the glass, the stopper POPS out.
      The heavier the mass of the stopper (rubber as compared to cork), ***the larger the recoil distance***, as the momentum ($M_1 V_1$) of the stopper is equal to the momentum of the test tube ($M_2 V_2$) in which M = mass and V = velocity. When only M_1 is increased, and all other variables are held constant, it will result in an increase of V_2.
      Using a wider test tube means increasing the stopper's mass, but also that of the test tube, thus the recoil would most likely stay the same. Using a longer test tube would change M_2 and thus the amount of recoil is not as large. This can be compared to using a revolver and a rifle (less recoil in a rifle, because of the increase of mass in using the rifle).

Invitations to Science Inquiry – Supplement – Page 138

SPACE SCIENCE	ANGULAR MOMENTUM
MOMENT OF INERTIA

THE SPINNING FOOTBALL

Materials: 1. An American football (actual or toy size).   2. A smooth table top.

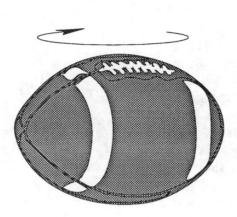 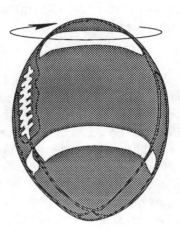

Procedure:
1. Place the football horizontally on the table top.
2. Hold it down firmly with the fingers of one hand and give it a quick spin.
   Observe! (If nothing happens, it was not spinning fast enough: try again).
3. Instead of a football, a hard-boiled egg may be used and it will behave the same way.

Questions:
1. What did you observe the ball doing?
2. What made the football spin vertically?
3. What would happen if you spin the football slowly (horizontally)?
4. What would happen if you spin the ball vertically?
5. Why doesn't the football stay spinning vertically?
6. What would happen if you spin the football fast vertically?
7. Why can't you spin the football fast horizontally and keep the ball spinning horizontally?

Explanation:

It takes more energy to rotate or spin the football horizontally as compared to vertically. This is because the mass of the football is spread out further from the center of gravity which is also the *center of rotation*, when it is in the horizontal position. There is a tendency for a system to move toward a *state of lowest internal energy or enthalpy,* which is the case when the football moves from the horizontal to the vertical position. But in order to spin at this position a certain energy level is required. This is why a slow spin at the horizontal position will just keep it spinning horizontally.

We find this principle applied in our daily life in figure skating, where the skater will spin faster when he/she pulls in his/her arms closer to the body while spinning. *(Conservation of Angular Momentum).*

During the high dive in the swimming sport, we see that divers keep their body closely tucked in when doing forward or backward saltos or flips. When they want to slow down the rotation, they stretch their body and this usually happens just before they enter the water.

Invitations to Science Inquiry – Supplement – Page 139

SPACE SCIENCE   ANGULAR MOMENTUM
                                                          PRECESSION

THE HUMAN GYROSCOPE

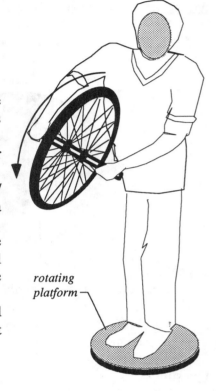

Materials: 1. An old bicycle wheel.
2. A wooden board (about 50 x50 cm), an old broom stick.
3. A hand full of steel ball bearings.

Procedure:
1. Cut a couple of pieces of broom handle and attach them to the axle on each side of the bicycle wheel for handles (drill holes in one end of the wood and screw the wood in the axle end).
2. Place the wooden board on the ball bearings on a smooth floor surface and have someone stand on the board.
3. Hand the wheel to the person standing on the board and let him/her hold the wheel vertically while you spin it with a downward motion.
4. Now ask the person holding the spinning wheel to rotate the plane of rotation from the vertical position to a horizontal position by turning it to the left. (What happens to the person?)
5. Now ask him/her to bring it back to the vertical position and keep turning it to the right to a horizontal position (What happens now?).

Questions:
1. Which way did the person holding the wheel turn, clockwise or counterclockwise (top view) when turning the spinning wheel to the left?
2. After turning the wheel to the left, in what direction does the wheel spin? Clockwise or counter-clockwise? (looking from the top).
3. After bringing the wheel back to its original vertical position, what happened to the person holding the spinning wheel?
4. In which directions did the wheel and person rotate by turning the spinning wheel to the right?

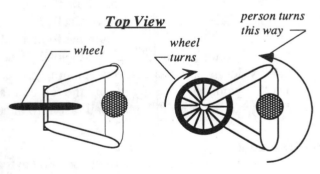

Explanation:
By turning the spinning wheel to the left, the wheel was turning in a clockwise direction and the person on the platform rotated counter-clockwise : *conservation of angular momentum.* The faster the wheel spins and also the more mass the wheel has (attaching lead weights to the wheel rim would be effective), the greater the angular momentum; thus the more difficult to turn the *plane of rotation* (flywheels in ships). The turning to the left of the person when the wheel is tipped to the left is called *precession*. This occurs f.i. in motorcycle riding: in order to turn left the motorcyclist tips the vehicle to the left and almost doesn't have to move his steering. A right turn is made by tipping the vehicle to the right.

SPACE SCIENCE   ANGULAR MOMENTUM
                                                              MOMENT OF INERTIA

THE TIN CAN RACE

Materials: 1. A variety of round tin cans of soup, vegetables, fruit cocktail, tomato juice, including solid dog food.   2. A stop watch (optional).
3. A one meter long inclined plane (a wooden board propped up at one end with a stack of books).

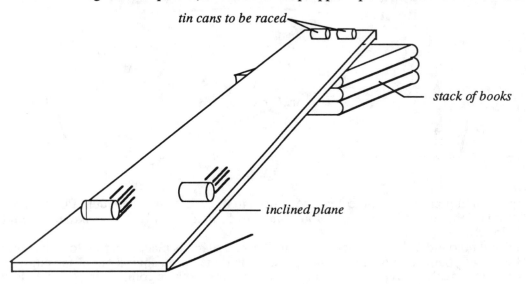

Procedure:
1. Set up the board as in the Sketch to form the inclined plane.
2. Divide the class in groups of at least three students in each group: one student to release the cans at the top of the plane, one to catch the cans at the bottam end of the plane, and one to record the time or to observe which of the two cans won the race.
3. Let each group of students choose a tin can that they think should roll the inclined plane the fastest and be the winning can (take average of three).
4. Let two groups at a time race their tin can against each other. The winner of this race will compete against the next group, etc. until all groups have participated. Declare the overall winner!

Questions:
1. Which of the tin cans is the overall winner?
2. What is the consistency of the contents of the winning can?
3. Haw did the tamato soup compare to the chunky soup?
4. How did the larger tin cans compare to the smaller ones?
5. Were the heavier cans always winning from the lighter ones?
6. What are the variables involved in determining the rolling speed?
7. Which are the manipulated and responding variables?
8. How can we control the other variables?

Explanation:
The variables involved in the tin can race are: rolling speed (responding), size, contents, weight of can, incline of the plane, etc. Any of the latter variables can be the manipulated one. Say we wanted to manipulate the size of the can, then we take two different sizes of cans containing the same thing. If we take, f.i. a can of soup and the solid dog food (with the same size), then we are manipulating the contents (in this case the consistency). Invariably, the can with the solid contents will win the race, as the ***moment of inertia*** is the smallest to overcome. Cans containing loose chunks of mass, like chunky soup or fruit cocktail are most likely the slowest, as they have the tendency to get the mass on the periphery of the can and thus the highest moment of inertia. (Hollow and solid cylinders can also be used to compare rolling speed).

Invitations to Science Inquiry – Supplement

SPACE SCIENCE   WEIGHTLESSNESS

THE CUP OF COFFEE DROP

Materials:
1. A styrofoam cup, a large wide and deep bucket.   2. Coffee or intensely colored water.

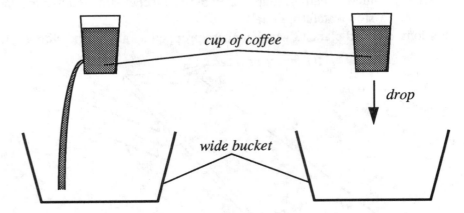

Procedure:
1. With a pencil point poke a hole near the bottom in the side of the foam cup. Place a finger over it and fill the cup with coffee or colored water (if colored water is used, make sure it is intensely colored).
2. Stand and hold the liquid-filled cup over the bucket, which is placed on the floor in front of you. Let go of your finger over the hole and show that the liquid squirts out of the cup.
3. Place your finger back over the hole and tell students that you are going to drop the whole cup. Let them follow the falling cup and observe closely whether any liquid is squirting out of the hole while the cup is falling. Now drop the cup and simultaneously remove your finger which was covering the hole.

Questions:
1. What makes the liquid squirt out of the hole in the beginning?
2. Would the strength of the squirt be the same on the surface of the moon, all other variables being equal?
3. Why did the liquid stop squirting during the fall?
4. What would happen if the finger covering the hole is removed a little sooner than the release of the cup?
5. If this same liquid-filled cup is held in an orbiting satellite, would any liquid squirt out of the hole? Would it be possible to hold the liquid in an open cup?

Explanation:
   When the liquid-filled cup is held stationary, the liquid is pulled down by the earth's gravity causing a liquid pressure at the point of the opening in the side of the cup, resulting in the squirt. When the cup is falling, the liquid pressure is suddenly eliminated, and no liquid is squirting out.
   All variables being equal, when this liquid-filled cup is held on the surface of the moon, the liquid will only squirt about one sixth of the distance from what it was on the earth, as the gravity pull is about 1/6th of the earth's.
   In a space satellite, no liquid will come out of the hole at all, just like in the falling cup, all materials actually are weightless. It would be very difficult to keep liquids in an open container in a satellite, this is why liquids are always stored in plastic bags.

SPACE SCIENCE   WEIGHTLESSNESS

THE WEIGHTLESS NAIL

Materials: 1. A coffee can plus lid.   2. A nail and thin string.

Procedure:
1. Make a small hole in the center of the coffee can lid.
2. Tie the nail to one end of the string (about .5 m) and run the other end through the hole in the coffe can lid.
3. While holding this end of the string, place the lid on the can.
4. Place the can on the table; hold the nail by the string and let go of the string: **CLICK!**
5. Now pick up the can and the nail by the string and hold them about 1 m (3 ft) above the floor or the table top.
6. Let students listen closely, and let go of the string: **CLICK - CLICK!**

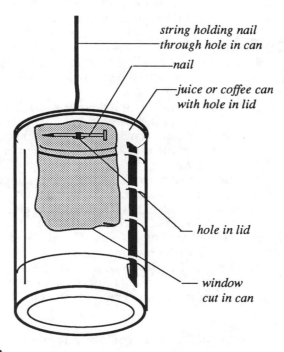

LOOKING UP ON CAN

Questions:
1. What made the first click? And the second click?
2. Where was the nail during the falling of the can?
3. In relation to the can, did the nail have any weight?
4. What other objects could we use in place of the nail?
5. When do we experience an almost weightless situation of our own body?
6. In a capsule that orbits around the earth, what can we compare to the can and the nail?
7. Would the two clicks always occur at the same time one after the other?
8. What characteristic of the tied object should be changed to make the span between clicks longer? And what property of the can itself?

Explanation: When coffee can and nail were released at the same time, they were falling at the same rate (disregarding the air resistance against the can). During this time, therefore, no sound was heard until the can hit the floor - first **CLICK**. The nail was still falling then and at the moment it hit the bottom of the can we heard the second **CLICK**.

The time span between the two clicks can be lengthened **only** when a longer can is used (dis regarding air friction). This will make the falling distance of the object longer and thus it increasing the falling time. Changing the weight of the falling object inside the can has no influence upon the rate of falling, since this rate is the same for all falling objects. When we take air friction of the falling object inside the can into consideration, a wider and flatter object would probably fall slower inside the can, and thus lengthen the time span between the two clicks.

This falling coffee can and the nail inside it may be compared with a capsule orbitting the earth and the people inside the capsule respectively. Since the capsule is constantly falling - *continuous freefall* and the people inside the capsule are falling with the same rate, the people inside the capsule experience weightlessness. Meaning that in relation to the capsule they are weightless. Actually all objects inside the capsule are weightless, but maintain their masses.

We can experience some weightlessness at certain rides in fairs: for example in the "Falling Drum" or the roller coaster, at the moment that the drum or car that's holding us almost falls freely. This we experience to a lesser extent in fast moving elevators. The sudden acceleration in good elevators are very much cushioned or more gradual, so that the rides become much smoother.

EARTH SCIENCE   SPECIFIC GRAVITY

WHAT IS THE ROCK'S S.G.?

Materials: 1. A small spring scale (one that can be hand held).
2. A container for water (large enough to contain the whole rock)
3. An irregular shaped rock.   4. Thin string.

Procedure:
1. Tie the rock on the end of the string and hang it from the spring scale (see Sketch A).
2. Read off how much it weighs and make a note of it. (reading A)
3. Fill the container with water, such that the whole rock can be immersed in the water.
4. Hang the rock from the scale and plunge it in the water (see Sketch B).
5. Read off how much it weighs now and make a note of it. (reading B)
6. Subtract reading B from A, and divide this difference into reading A.
7. The specific gravity or density is expressed in grams (mass units) over cubic centimeters (volume units).

Questions:
1. Which of the two readings was greater?
2. Did the water level in the container go up or down after immersing the rock in the water?
3. In what ways can we measure this increase in water volume?
4. What property of the rock is this volume difference of water equal to?
5. Would two different types of rocks with the same mass displace the same amount of water?
6. Would two rocks which displaced the same amount of water have to be of the same type?
7. What other properties do we have to look at besides the *specific gravity,* tto determine whether rocks are exactly of the same type or not?

Explanation:
Since the difference in weight of the rock in air and immersed under water is equal to the upwards force or *boyancy force,* it is also equal to the displaced water volume or weight (since the density of water is 1). This displaced water can be measured in a graduated measuring cylinder or in an over flow can (see Event 15.7 ISI-2nd Ed). This volume is important, because it is exactly equal to the volume of the rock itself. As the mass of the rock is know from the first weighing (reading A) and the volume of the rock is calculated from the difference of the two weighings (reading A - reading B), the *specific gravity* or *density* of the rock can be calculated from A/(A-B).
 Different rocks of different types would have different specific gravities. This is only oneproperty of rocks that we would look at in order to determine whether the rocks are of the same type. The more obvious properties of rocks are of course their color and hardness. This last property is usually checked by scratching the rock with the human nail (No 4 of the hardness scale: from 1, softest, Talcum to 10, hardest, Diamond). Another property is the crystalline structure, usually quickly checked by breaking the rock and examining the cleavage.

Invitations to Science Inquiry – Supplement –

EARTH SCIENCE   ROCKS & DENSITY

HOW CAN WE DETERMINE THE ROCK'S VOLUME?

Materials: 1. A medium size tin can. 2. A measuring cylinder, a piece of thread.
3. Different rocks, including: pumice, sandstone, granite.

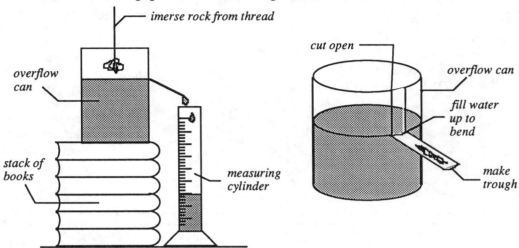

Procedure:
1. Make an overflow can by cutting a 2 cm vertical strip of about 5 cm long out of the side of the can and bend this strip outward, making a small gutter out of the strip by bending the outer edges up (see Sketch).
2. Place this overflow can on a stack of books and fill it with water until the water just overflows.
3. Place the measuring cylinder underneath the end of the overflow trough to catch the overflowing water.
4. Tie the rock to be measured to a string and lower it slowly into the overflow can until it is completely submerged. Read off the volume of the water in the cylinder (for floating objects, like pumice, use a needle to immerse the whole object underneath the surface of the water).
5. Weigh each of the rocks on a scale and let students determine the densities of each of the rocks (D = M / V).

Questions:
1. What made the water overflow?
2. What influences the amount of water overflowing?
3. Why does the pumice stone have to be pushed under water?
4. What result would we get if the pumice was not completely submerged?
5. What are the variables to be controlled if we wanted to compare the water displaced by the different rocks?
6. Does the shape of the rocks influence the amount of overflow?
7. Would a round and a cubical rock of the same type and weight (mass) have different amounts of overflow?

Explanation:
The different types of rocks have different densities, and as *density is defined as being mass of the object per unit of volume*, it is necessary to determine the volume and the mass of each of the rocks. The volume of an odd shaped object (which rocks usually are) can easily be determined by submerging it in water and measuring the displaced water. This can be done with the overflow can.

When rocks of the same weight but of different types are submerged in the overflow can, different volumes of displaced water will be obtained. This is caused by the different densities of the rocks. The amount of overflow for the same type of rock will be the same, as long as the mass of the rock is held constant (the same), no matter what shape it is in.

Invitations to Science Inquiry – Supplement – Page 145

SIMULATE A VOLCANO ERUPTION

Materials: 1. Ordinary earth clay & a wooden board (about 1m x 1m).
2. Ammonium bichromate (50 g), magnesium powder (12 g).
3. Magnesium ribbon (about 1 m long).   4. A wooden broom stick.

CAUTION:

PERFORM OUTDOORS OR UNDER FUMEHOOD!

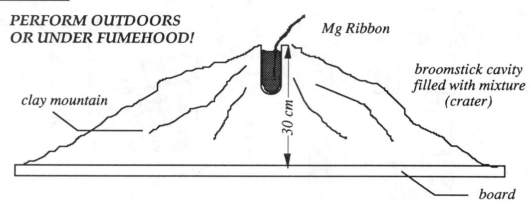

Procedure:
1. With the wood board as a base, have the students build a volcanic cone with the clay of about 30 cm high and a diameter of 60 cm at the base. Push the handle of a broom in the middle of the volcano down to about 6 cm deep to form the crater
2. The 50 g ammonium bichromate is enough for about 3-4 eruptions. Mix about one third of the amount of bichromate with about 4 g of magnesium powder and pour the mixture in the crater.
3. Cut a 10 cm strip of magnesium ribbon and push one end into the mixture filling the crater, let the other end stick out of the crater for a fuse.
4. Light the magnesium ribbon fuse with a match and stand back! (If the eruption did not set off, wait a few moments, then insert a second fuse and try again).
5. After the eruption, a second eruption can be simulated by pouring more of the bichromate-magnesium powder mixture into the crater while it is still hot (use a long paper or cardboard chute to fill crater).

Questions:
1. What is the function of the magnesium ribbon?
2. What was the cause of the eruption (as a chemical reaction)?
3. What causes an actual volcano to erupt?
4. What comes out of the crater of an actual volcano?
5. How can actual volcanic eruptions be predicted?

Explanation:
Ammonium bichromate $(NH_4)_2Cr_2O_7$ is a very strong oxidant and unstable when heated. At the time that the magnesium ribbon flare touches the mixture of bichromate and magnesium powder, the bichromate decomposes into ammonia gas, chromium oxide, and oxygen. In the presence of the magnesium powder it also forms magnesium chromate. The sudden formation of the gases is the main cause for the upward spray and spewing out of the chemicals out of the crater.

A real volcano erupts because of the pressure built up in the earth. The craters in the volcanoes are acting as valves through which this earth's pressure is released. When erupting, a volcano spews hot molten earth, called *magma* (underneath the earth's surface) or *lava* (above the earth's surface) out of the crater.

PLANTS   GROWTH
                                                                 OSMOSIS

THE SWOLLEN EGG

Materials: 1. A fresh chicken egg.   2. Dilute HCl or strong vinegar, a cup or beaker.
3. An old fashioned milk bottle (or wine carafe or any bottle of which the mouth diameter is a bit smaller than the egg's).

Procedure:

1. Place the egg in the cup and add dilute HCl (hydrochloric acid) or strong vinegar to it until it is completely immersed. Hold the egg under the surface of the liquid or keep rotating the egg. Do this for 10 minutes.
2. Check whether all the egg shell has dissolved by touching the egg and very carefully pressing on it. When only the membrane is left, it should feel soft and flexible.
3. Fill the milk bottle about 3/4 full with hot water (almost boiling) and place the egg immediately on the mouth of the bottle and let stand.
4. Observe! Leave for an hour or so and observe again!

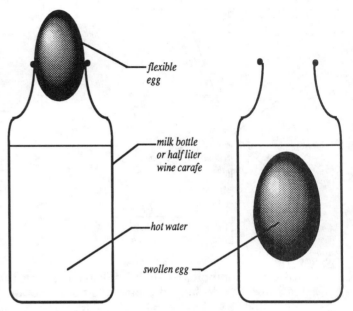

Questions:

1. why did the egg become soft and flexible after placing it in acid?
2. What did the hot water do to the air in the milk bottle?
3. What did the flexible egg do after it was placed on the bottle?
4. What did the egg do after an hour wait?
5. What made the egg grow larger in the bottle?
6. How can we get the egg out of the bottle without breaking it?
7. How can we reverse the process of osmosis?

Explanation:

The shell of an egg consists mainly of calcium carbonate. By placing the egg in dilute hydrochloric acid or strong vinegar, the calcium carbonate reacts with the acid to form a soluble calcium salt and carbondioxide gas. This leaves only the membrane around the egg, which makes the egg feel soft and flexible.

The steam of the hot water drives the air partly out of the bottle, thus by placing the flexible egg on the bottle, the egg is slowly sucked into the bottle while the water is cooling off.

After the egg is totally plunged in the water, *osmosis* is taking place. Water molecules are migrating through the ***semi-permeable membrane*** of the egg from the outside into the egg and thus making the egg swell. In order to get the egg back out, we have to shrink the egg, which can be done by pouring all the water out and letting the egg dry out and shrink. Then by heating the bottle while the egg sits in the neck, it will slowly be pushed out by the expanding air.

PLANTS   CAPILLARY ACTION
                                                   ABSORPTION

THE TOOTHPICK STAR

Materials: 1. Five toothpicks (flat kind), 1 box for the whole class of 30.   2. A cup of water.

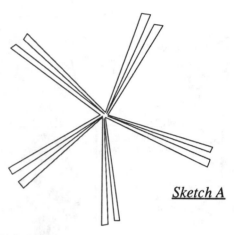
Sketch A

Sketch B

Procedure:
1. Break five toothpicks in half but leave the parts together, and leave them in a V-shape.
2. Place the five V-shaped toothpicks on a smooth surface with their points as close as possible together (see Sketch A).
3. Dip your finger tip in the cup of water and let one drop fall in the center of the toothpick configuration.
4. Observe carefully what is happening to the toothpicks!

Questions:
1. What happened to the toothpicks?
2. What did the water do with the toothpicks?
3. What material do the toothpicks consist of?
4. What will happen to picked flowers when they are not placed in a vase in water? Why do they need water?
5. How does water reach the top leaves in a tall tree?
6. Compare a dry broken toothpick with one that was wetted by holding one of the legs of the V-shape and pushing the other leg down: Which one stays closed/down?

Explanation:
By breaking the toothpick in half, not all the vessels/fibers are cut off. The water is *absorbed* by the wood fibers by *capillary action,* which consists mostly of adhesive forces. Also *diffusion* through the fiber walls is taking place: water is thus filling the unbroken vessels or fibers. This makes the toothpick tend to straighten up and make it more springy than the dry one.

This is the reason why picked flowers become limp when they are not placed in water. Capillary action and *osmotic pressure* are the forces that bring the water from the roots of a tree all the way up to the top leaves.

When repeating the activity, dry toothpicks have to be used. The wet ones will not form and stay in a narrow V-shape. They become more springy than the dry toothpicks and form a much wider V-shape.

Invitations to Science Inquiry – Supplement – Page 148

HUMAN BIOLOGY   THE HUMAN EYE

ARE WE PARTIALLY BLIND?

Materials: 1. A blank sheet of paper.   2. A pencil or pen.

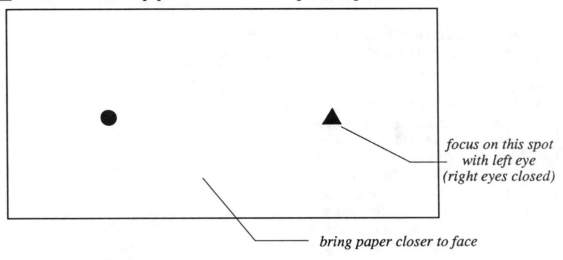

focus on this spot with left eye (right eyes closed)

bring paper closer to face

Procedure:
1. Draw a small circle (about .5 cm diameter) and a small triangle (about the same size), about 15 cm apart on the white blank sheet, and completely darken them in.
2. Hold the paper about 30 cm in front of your eyes and shorten this distance very gradually (bring the paper closer to you slowly). While doing this look with your left eye (close right eye) and focus on the triangle or look with your right eye (left eye closed) and focus on the circle.
3. If you do not notice anything happening with the other dot, you may want to hold the paper somewhat higher or somewhat lower while approaching it.

Questions:
1. Why does the circle disappear at a certain distance from your left eye?
2. Is the angle or the distance between the circle and the left eye more important in getting it to disappear?
3. Does it make any difference if we had other shapes on the paper?
4. What makes us see the circle although we're focussing on the triangle?
5. How does the human eye function?
6. Is the retina of our eye totally complete and continuous?

Explanation:
The retina of the human eye may be compared with the photographic film in a camera. The big difference is that in the human eye it is nerve cells that send electrical impulses to the brain when light rays fall on the retina. These nerves are bunched up and coming into the retina at a certain spot, called: *the blind spot,* and it is at this spot that very small images are undetectable (see Sketch on right).

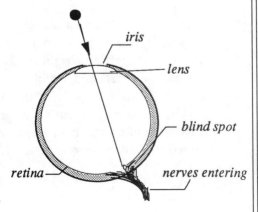

Top view of left eyeball

HUMAN BIOLOGY   EYESIGHT

ARE YOU LEFT- OR RIGHT-SIGHTED? (II)

Materials: 1. A pencil or straw or ruler (or any straight edge).
2. A vertical straight edge of the room (inside/outside corners).

Procedure:
1. Hold the pencil (or straw, or any other straight edge) with an outstretched arm vertically in front of you. (if no straight edge is available you can use your index finger and hold it vertically up).
2. Find a straight vertical edge (corner) of the room (or any other straight line) and align the pencil with it, holding both eyes open.
3. While keeping the pencil aligned with the straight edge of the room, close one eye at a time. Which eye do you close to make it look like the pencil jumps aside?

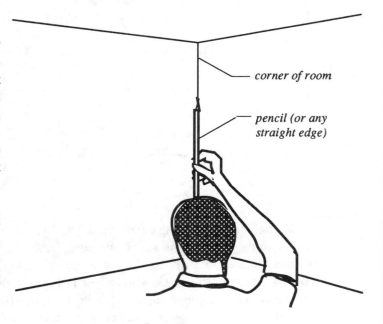

Questions:
1. Does the pencil jump aside when closing your left or right eye?
2. If the pencil jumps aside when closing your right eye, are you using more your left or your right eye? In other words, is your left or right eye the dominant one?
3. If the pencil jumps aside when closing your left eye, are you left-sighted or right-sighted?
4. What makes the pencil jump aside when one of the eyes is closed?

Explanation:
For the majority of people the pencil or whatever straight edge is aligned with the vertical line of the room, will jump aside when the right eye is closed. This indicates that most people are *right-sighted*. The reason for the pencil to jump aside, is that the right-sighted person uses his/her right eye more than the left. (the right eye is the *dominant* one). Thus the pencil is lined up in a straight line between the straight edge of the room and the right eye in the first place. When this eye is closed, the left eye still sees the pencil, but it is not in line with the edge of the room.

This is one of the easiest ways to tell whether a person is left-or right-sighted. If a pencil is not available, a forefinger held vertically may be used just the same. Left-handed people usually are left-sighted or in other words: their left eye is usually the dominant one.

Invitations to Science Inquiry – Supplement – Page 150

HUMAN BIOLOGY   PERCEPTION

THE SWAYING CARDBOARD

Materials: 1. A white 5 x 8" cardboard or heavy paper.

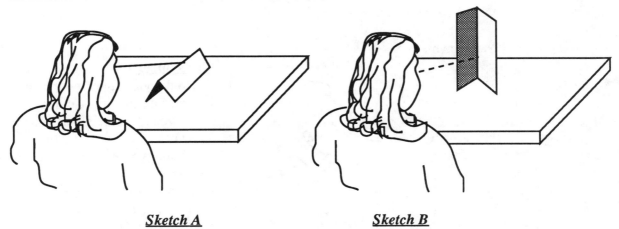

Sketch A   *Sketch B*

Procedure:
1. Fold the paper card along the long axis in half.
2. Place the folded card directly in front of you on the table with the ridge pointing in your direction (See Sketch A).
3. Select a spot in the center of the fold and stare at it steadily with one eye shut (if you're right-sighted cover your left with your hand).
4. First you will see the card like in Sketch A. Continue staring at it until suddenly you see the card in its second position (like in Sketch B).
5. When you see the card in its second position (standing up), move your head slowly from side to side (still keeping one eye shut). Do you observe the card swaying back and forth?

Questions:
1. How long did it take you to see the card in its standing position?
2. What did you see the card do when moving your head from side to side (with one eye shut)?
3. When seeing the card in its second position, are all the depth information and the way the light and shadow falls on paper and table, consistent?
4. What do you see when moving your head closer and farther from the card?

Explanation:
This interesting double illusion demonstrates a phenomenon called *parallax*. When first seeing the folded card, it is lying down and you see it lying in the position as in Sketch A. All the information of the depth and how the light and shadows are falling, are consistent. When seeing it in the second position (like in Sketch B), the cues are all contradictory: the near point of the fold becomes the farthest point and the far corners of the paper become closest. Normally, when moving your head from side to side, near points will move more than far points of an object (try it with both eyes open). Seeing the card in its second position reverses these points and thus gives us an unusual distorted image.

HUMAN BIOLOGY   PERCEPTION
                                                  EYESIGHT

THE FLOATING PIECE OF FINGER

Materials: 1. Yourself and your two index fingers.

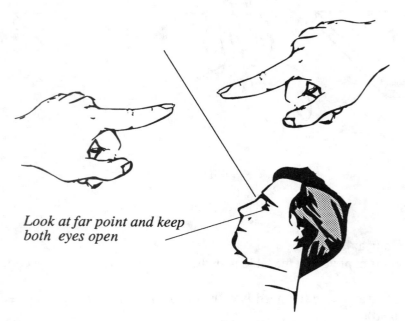

Look at far point and keep both eyes open

Procedure:
1. Hold your left and your right forefinger about 30 cm in front of you at the height of your eyes. Hold them horizontally about two-three cm apart (see Sketch on left).
2. Do not focus your eyes on the fingers but look over them and focus at a far point.
3. Wiggle the fingers slightly up and down. Can you see a double-nailed piece of finger floating in the air?! (Do not focus on your fingers!)

Questions:
1. What made you see only a piece of each finger?
2. What happens if you focus your eyes on the fingers?
3. What happens if you look under the fingers and focus at a far point?
4. What do you see if you put your thumbs in the same position?
5. How can you check what your left eye is actually seeing?
6. Try closing one eye at a time; what do you see?

Explanation:

Even though the eyes are focussed on a far point, we still see objects that are closer to the eyes. The image from the left eye and the image projected in the right eye are both combined in our brain.

This is the reason why we see only a piece of the finger. Whatever

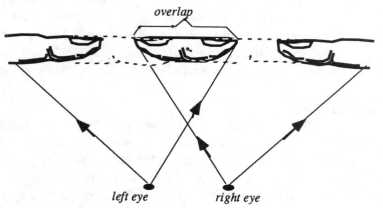

image overlaps is seen more clearly. You would never be able to see the same floating piece of finger with one eye closed. We are not able to see depth with only one eye. It is like looking at a two dimensional picture, if we use only one eye.

Invitations to Science Inquiry – Supplement – Page 152

HUMAN BIOLOGY   ILLUSION
                                                                 EYESIGHT

THE HAND IS QUICKER THAN THE EYE

Materials: 1. Two quarters or two other identical coins.

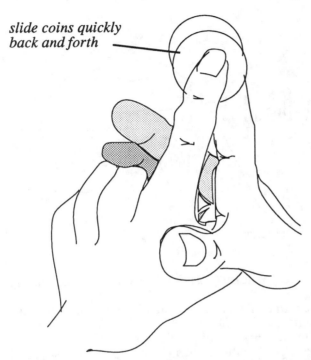

slide coins quickly back and forth

Procedure:
1. Place two coins between your two index fingers (conceal this from the students or audience, do not tell them how many coins you have).
2. Then rub the two coins back and forth as quickly as you can.
3. While rubbing, show them to the students (close up), and ask: "How many coins do you see?" Anticipated answer: "Three". If the answer is "Two", let them take a good look again at the moving coins.
4. After the majority of students (ideally all students) say that there are three coins between your fingers, go to the student that said "definitely three", and drop the two coins in his/her hand.

Questions:
1. How many coins were actually moving?
2. Why do we see three coins when the coins are rubbed against each other?
3. What does it mean: the hand is quicker than the eye?
4. What do you see when the rubbing is considerably slowed down?
5. What are daily applications of this same principle?

Explanation:
The human eye is a marvelous instrument. It operates on the same principle as the camera. The eye, however, is able to take two simultaneous pictures, one in black and white and the other in color. Cells called *rods* in the retina register black and white only; and other cells called *cones* register the different colors. The retinal cells are so sensitive that they can detect light as feeble as a 100-trillionth of a Watt. This is 1×10^{-11} Watt. The retina also has the capacity to *retain an image* a little shorter than half a second, this means that if the coins are moved faster than half a second per position, the eye still sees the coin at the original spot. Thus it sees three coins instead of two.

Other applications of this same principle are: motion pictures, animated cartoons, spokes of a moving wheel that sometimes are seen moving in reverse, etc.

Invitations to Science Inquiry – Supplement – Page 153

HUMAN BIOLOGY   ILLUSION
                                                                 EYESIGHT

PUT THE BIRD IN THE CAGE

Materials: 1. A half dollar or dollar coin (a quarter is OK too).   2. Two large sewing needles.

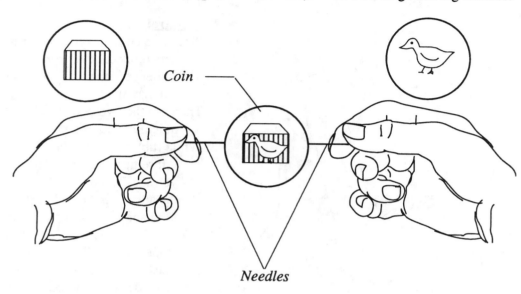

Procedure:
1. Cover both sides of the coin with tape that you can write on (masking or 3M magic tape) and trim the edges off.
2. First draw a cage on one side, then you draw a bird on the other side, making sure that you flip the coin upside down, and not sideways, before drawing the bird.
3. Show the drawings to the audience and ask them: "How can I put the bird in the cage?"
4. Find the diameter of the coin half way up the cage by cutting out a piece of paper the size of the coin and fold it in half.
5. Place the coin on top of the paper circle and pick the coin up with the two needles exactly along the diameter of the coin (see Sketch).
6. With the two needles bring the coin close to your mouth and blow against the upper half of the coin.
7. Repeat the blowing with short puffs. Observe the spinning coin: Bird in the cage!

Questions:
1. Why do we place the two needles along a diameter of the coin?
2. What would happen if they were not placed along a diameter?
3. Around what point do spinning objects rotate?
4. What made us see the bird in the cage?
5. Did the bird actually go in the cage?

Explanation:
　　All spinning objects spin around their center of gravity. This is why it is necessary to find a true diameter of the coin and place the two needles along this diameter. If the needles were not placed along a diameter, the spinning would not go very smoothly.
　　Seeing the bird in the cage is only an ***illusion,*** just like in moving pictures. The image of the cage or the bird is retained longer by our eyes than the time it takes for the coin to flip around. Thus both images are combined by our brain and we see the bird in the cage.

HUMAN BIOLOGY

ILLUSION
EYESIGHT

THE ELLIPTICAL PENDULUM SWING

Materials: 1. A thin string, a washer or nut (or other bob).
2. A polaroid filter, or one lens of a pair of sun glasses, or semi-dark film negative.

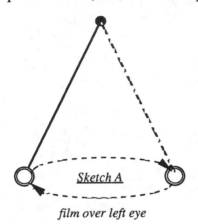
Sketch A
film over left eye

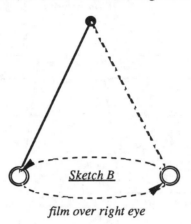
Sketch B
film over right eye

Procedure:
1. Attach the washer or metal nut to the thin string to make a pendulum.
2. Let the pendulum swing from a fixed point (someone could hold the end of the pendulum with a steady hand).
3. Look at the swinging washer with both eyes open, but cover your left eye with the polaroid filter (or other screens). What do you observe?
4. Now look at the swinging pendulum with the film over your right eye. What do you observe? In which direction is the bob swinging?

Questions:
1. What is the reason for seeing the pendulum swinging in an eliptical path?
2. Would it be easier to see objects in a dark or in a bright room?
3. Would the eye perceive an object faster or slower if that object is brightly or less brightly illuminated?
4. If one eye is perceiving a moving object a fraction of a second slower, where would this eye actually see the object? In front or behind the actual position of the object?

Explanation:

The eye that is covered by the darkened filter perceives the pendulum *a fraction of a second later* than the uncovered eye. This means that the left eye (say that this eye was covered with the filter) perceives the washer a little behind the actual position A (see Sketch below). Our brain interprets these two perceptions as coming from one object and combines the two images, thus we see it either a little farther or a little closer than the actual position.

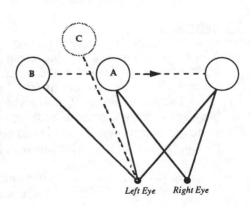

The Sketch on the right illustrates the washer A moving to the right. The left eye (covered) perceives it at position B and the brain interprets it in position C. On the way back the brain interprets the position of the washer as being closer than the actual spot. Thus the motion is perceived as being eliptical.

Invitations to Science Inquiry – Supplement – Page 155

HUMAN BIOLOGY   NERVOUS SYSTEM
                                                              TOUCH

IS THE WATER WARM OR COLD?

Materials: 1. Three beakers (400 ml) or large cups.

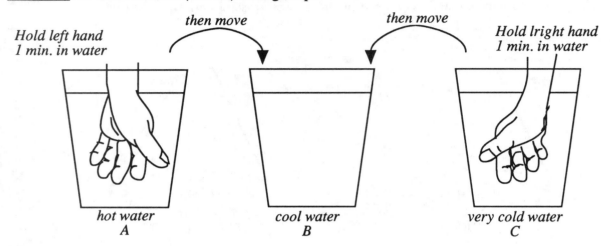

Hold left hand 1 min. in water — then move → hot water A — then move → cool water B — Hold lright hand 1 min. in water — very cold water C

Procedure:
1. Fill a beaker (A) three quarters full with hot water (about 50°C) and another beaker (C) three quarters full with very cold water (about 5°C) (add a few ice cubes if water is not cold enough in the summer time).
2. Fill the third beaker (B) with regular water of room temperature (20°C) and place it in between the hot and the cold water beaker.
3. Immerse your left hand in the hot water and hold it under water for one minute, then move it in the center beaker. Does the water feel warm?
4. Immerse your right hand in the cold water and hold it under water for one minute, then move it in the center beaker. Does the water feel warm or cold?

Questions:
1. Did the water in the center beaker (B) change in temperature?
2. Why did the water in beaker B feel warm with one hand and cold with the other?
3. What gives us the sensation of warm or cold in our body?
4. What are other similar situations where this same principle applies?

Explanation:
When the left hand is placed in the hot water, the ***nerve endings*** in our nervous system send a message to our brain, telling it that it feels warm. Over time, however, the nerve endings are dulled and the hand adapts itself to the warm sensation. When it is placed into the water of room temperature, there is a difference of about 30°C lower, and the water suddenly feels cold to this hand.

The same interpretation can be put forth for the hand immersed in the cold water. This time it's only going the opposite way: from cold to warm. Similar situations are encountered when we take a shower with very cold hands and adjusting the water temperature with the touch of our cold hands; placing our body under the shower, the water still feels cold because to the hand the shower water already felt very warm.

Someone used to living in a home with 18°C room temperature might find on visiting a friend's house with a 22°C room temperature quite warm, and of course visa versa.

HUMAN BIOLOGY   NERVOUS SYSTEM
                                                              REACTION TIME

CATCH THE DOLLAR BILL

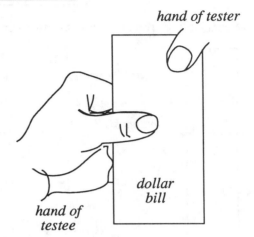

hand of tester

dollar bill

hand of testee

Materials: 1. A crisp dollar bill (any denomination)
(ideally: one of each denomination up to $20).

Procedure:
1. Hold your forefinger and thumb of your left hand about 5 cm (2 inches) apart, place the dollar bill about half way in between the two fingers.
2. Let the dollar bill fall and show the audience that you can easily catch it between your two fingers.
3. Ask the members of the audience to hold their forefinger and thumb in the same position to try to catch the dollar bill.
4. Hold the dollar bill between their fingers. The rule is that they may not go down with their hand, and the gap between the fingers should not be smaller than 5 cm.
5. Vary the time between placing the dollar bill between their fingers and the moment that you let go of the bill. Tell the audience that whoever catches the bill may keep it. Continue doing it with larger denominations, but remember to vary the dropping time, otherwise the bill might get caught by anticipation.

Questions:
1. Why can you catch the dollar bill so easily when you drop it yourself?
2. Why do you need to vary the time between placing the dollar bill and the releasing of it?
3. What is the reason that the bill can never be caught?
4. What makes muscles contract?
5. How did the finger muscles get the order to catch the bill?
6. Where did the initial stimulus came from?

Explanation:
The initial *stimulus* to catch the dollar bill was from seeing the falling of it. The eyes got the stimulus, they sent a signal (*response*) to the brain and the brain in turn sent an electrical impulse to the finger muscles to contract and catch the bill. But all that took a longer time than the falling of the dollar bill, although both take only a fraction of a second.

If you place the dollar bill between your left fingers and let it fall with your right hand, it is very easy to catch. This is because the stimulus or signal was already in your brain, and all it had to do is send the signal to the muscle of your fingers. The difference in doing it to yourself or having someone else do it to you, is the time it took for the light to travel from the dollar bill to the eye, and for the eye to send a signal to the brain.

The bill can only be caught by someone who anticipates the falling and actually start closing the fingers before the bill falls. This is the reason why we should vary the time between placing the bill between the fingers and the actual releasing of it.

HUMAN BIOLOGY

NERVOUS SYSTEM
REACTION TIME

HOW FAST CAN YOU REACT?

Materials: 1. A meter stick (15 meter sticks for a class of 30).

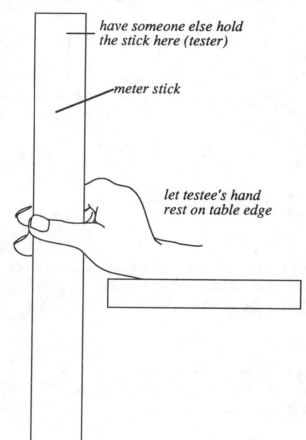

have someone else hold the stick here (tester)

meter stick

let testee's hand rest on table edge

Procedure:
1. Place your hand over the edge of a table and leave about 3 cm opening between the thumb and forefingers.
2. Let your tester (someone else) hold the meter stick vertically from the top end of the stick.
3. Read off the spot where your thumb is on the meter stick before the tester drops the stick (start at an even number f.i. 10 or 20).
4. Let the tester drop the stick. The moment that you see the stick drop catch the stick immediately and do another reading of where your thumb ended up. Record the difference (distance in drop).
5. To test the other stimuli: <u>Hearing</u>: close your eyes and let tester say: "Now" exactly at the moment that he/she drops the stick. Testing for <u>tactile (touch)</u>: close your eyes and let tester touch your other hand exactly at the moment that he/she let go of the stick.

Questions:
1. How do girls compare to boys in reaction time?
2. In doing the above comparison, what variables have to be controlled?
3. How do the different stimuli compare in your reaction time?
4. What is the actual falling time for the distance in the drop?
5. What are all the variables involved in this experiment?

Explanation:

The different variables in this experiment are: the *manipulated variable* is the *stimulus*: through the eyes, hearing, and touch (or sex: when comparing boys versus girls). The *responding variable* is the falling time or distance on the meter stick.

As the distance in free fall is: $d = 1/2\, gt^2$, the time derived from that is t = square root of $2d$ over g, where g is the gravity acceleration.

All other variables, like: distance between fingers, the place where the tester holds the stick, catching with two fingers or with the whole hand, the accuracy of saying "now" and simultaneously letting go of the stick, etc. have to be held constant, in other words they have to be done the same way.

Invitations to Science Inquiry – Supplement – Page 158

HUMAN BIOLOGY   MUSCLE COORDINATION

THE UNCONTROLLABLE FOOT

Materials: 1. A blank sheet of paper. 2. A pencil or pen.

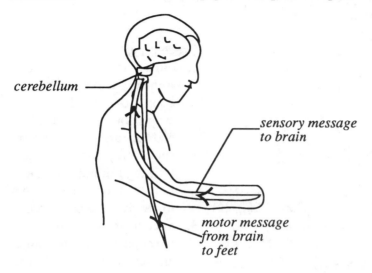

Procedure:
1. Stand up next to the table with the paper in front of you.
2. Hold the pen or pencil in your right hand (if you're right-handed).
3. If you're right-handed, let your right foot rotate in a clockwise direction (describing small circles on the floor with you right foot).
4. Try to keep your foot rotating while simultaneously writing a large *number 6* on the paper. Observe what your foot is doing!
5. Now repeat point 3 and 4 but with your left foot making the clockwise motions. Try drawing the six now with your right foot making counterclockwise motions. Then do the same with your left foot. What are your findings now?

Questions:
1. Which of the four combinations was easiest to carry out?
2. When moving your foot in a clockwise direction, was it easier to do it with your left or right foot (while drawing the 6)?
3. Did you try to write the sixes with your left hand (if you re right-handed) while rotating your foot?
4. What makes it so difficult to rotate your right foot (when right-handed) or your left foot (when left-handed) in a clockwise direction?
5. What part of the human body controls the muscles?

Explanation:
The *cerebellum* located in the base of the *cerebrum* of the human brain controls the coordination of voluntary movements. The cerebellum is important for such activities as walking, dancing, playing ball, or even for such routine tasks as tying a shoelace or writing a figure 6.

Doctors have known for generations that nerve fibres from the right side of the body cross over in the brain stem to the left side of the brain. Similarly, nerve fibres from the left side of the body cross over to the right side of the brain. In other words, the whole left side of the body is controlled by the right half of the brain, and similarly, the whole right side of the body is controlled by the left half of the brain.

By writing a figure 6 with the right hand, the left half of the brain has instructed the right hand to make a counter-clockwise motion. The right foot could easily make the same counter-clockwise motion, but the opposite movement requires a special effort. With practice both movements can be achieved. Drummers have no trouble performing this, because their muscles have been trained to do independent movements!

Invitations to Science Inquiry – Supplement – Page 159

HUMAN BIOLOGY   MUSCLE COORDINATION

THE KICKING FROG LEG

Materials: 1. A freshly cut frog leg. 2. A copper (bronze) bolt, and two nuts.
3. An iron stand or frame (galvanized iron strip will do).

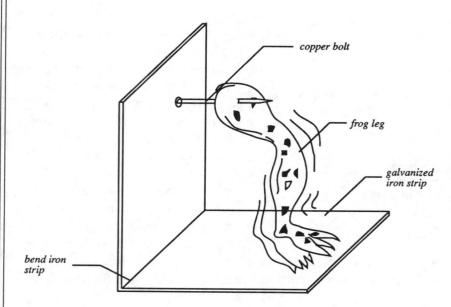

Procedure:
1. Drill a hole in one end of the iron strip and attach the copper bolt tightly to this strip with the two nuts.
2. Bend the iron strip at a spot in such a way that when the frog leg is hung from the bolt, the foot will touch the lower part of the iron strip.
3. Pierce the frog leg through the copper bolt in the thigh and let the leg droop down. Observe the spasms!

Questions:
1. What made the frog leg go into spasms?
2. What would be necessary for a muscle to contract?
3. Would an iron bolt do the same thing for the frog leg?
4. Would this demonstration work if you use an iron bolt in a copper stand?
5. What other metals would give this same contraction in the frog leg?
6. How can we find out which metals give the potential difference?
7. How can we compare this with human muscles and their contraction?
8. Why is it dangerous to grab a live wire on purpose?
9. How should we test a live wire with our hands?

Explanation:

There is a potential difference between copper and iron. At the moment that the frog leg touches the iron frame, while it is suspended from the copper bolt, an electrical current (a flow of electrons) is running through the muscle and makes it to contract. After a while the leg muscles relax and the leg droops down, the foot touches the iron and the leg goes into spasms again.

The further apart two metals are in the *electromotive series,* the larger the *potential difference*, in other words the larger the difference in tendencies to go into solution or give off electrons. Other combinations are: copper and zinc, magnesium and copper, iron and silver, etc.

Human muscles are also controlled by electric currents. It is therefore very dangerous to deliberately grab a "live" wire, as you may not be able to let go of the wire. In order to test a live wire, use the back of your fingers or hand and touch it for a short moment. If the wire sends electricity through your muscle, your fingers will be moving away from the wire rather than clamping around it tighter, if you touch it with the inside of your hand.

HUMAN BIOLOGY   BODY BUILD OF
                                                                MEN VS WOMEN

ARE WOMEN MORE AGILE THEN MEN?

Materials: 1. A plastic cup (or any other object of the same dimensions).

Procedure:
1. Let one of the ladies in the audience sit on her knees and lower leg (like in a kneeling position) on the floor.
2. Place a cup (or other small object) in front of her the length of her forearm away from her knees on the floor (see Sketch).
3. With arms behind her back, let her bend forward and knock the object over with her nose! Men cannot do this without falling forward! Woman can!

Questions:
1. Why can women knock the object over without falling, but not men?
2. Are women actually more agile and lean compared to men?
3. What is different in the basic skeleton built of women compared to men?
4. Where is the center of gravity of women located?
5. Where is the center of gravity of men located as compared to women?
6. How would the exceptions of women that cannot do the trick be built?
7. How would the exceptions of men that can do the trick be built?

Explanation:
In general, women's body structure is such that they have a lower *center of gravity,* because of their wider hips and heavier bone structure in the lower abdomen part of the skeleton as compared to men's structure. Similarly we can say for men in general, that they have wider shoulders as compared to women. This makes the center of gravity of men's bodies higher than women's.

This lower center of gravity is the main cause for women to be able to bend forward in the kneeling position without falling over forward. When men try to bend over they will fall forward because of the location of their center of gravity. When this center of gravity passes beyond the knees, their bodies will topple forward. (See also Event 13.10 of **ISI-2nd Ed** for a similar activity).

Index

A

ABSORPTION
   Surface Tension *43*
Action is Reaction *134*
Activation Energy *56*
ADHESION
   Center of Gravity
      The Center-Seeking Paper *103*
Adhesion *45*
ADHESIVE FORCES
   Friction
      Magic Strip of Newspaper *108*
Adhesive Forces *103*
AIR
   Aerodynamics
      Air Bullet *28*
      Smoke Ring Race *29*
   Exerts Pressure
      Balloon in Jar *25*
      Crushing Pop Can *26*
      Weightless Water *24*
   Pressure
      Water to Lemonade *27*
Air
   Condensed *25*
Air Bullet *28*
AIR RESISTANCE
   Center of Gravity
      Stand a Dollar Bill on Your Finger *105*
Air Resistance *105*
Alarm Clock *96*
Alka Seltzer *53*
Aluminum Foil *87*
Ammonia and Hydrochloric acid *28*
Ammonium Bichromate *146*
Ammonium Chloride *28*
Ammonium Phosphate *74*
Ammonium Thiocyanide *70*
Amplitude *79*
Angle of Incidence *92*, *97*
Angle of Reflection *97*
ANGULAR MOMENTUM *139*, *140*
   The Tin Can Race *141*
Angular Momentum *34*
ANOTHER BALANCING ACT *131*
Aquarium *89*
ARE WE PARTIALLY BLIND? *149*
ARE WOMEN MORE AGILE THEN MEN? *161*
ARE YOU LEFT- OR RIGHT-SIGHTED? (II) *150*

B

Barium Chloride *58*
Balancing Point *99*
Ball Bounce *137*
Balloon *25*, *59*, *69*, *87*
Barium Hydroxide *70*
Barometric Pressure *34*
Battery *87*
Bernoulli s Principle *31*
Bicycle Wheel *140*
BIOLOGY-HUMAN
   Body Build of Men vs Women
      Are Women More Agile Than Men? *161*
   Eyesight *150*
   Illusion
      Bird in Cage *154*
      Elliptical Pendulum Swing *155*
      The Hand Quicker Than The Eye *153*
   MUscle Coordination
      Kicking Frog Leg *160*
   Muscle Coordination
      The Uncontrollable Foot *159*
   Nervous System
      Catch The Dollar Bill *157*
      How Fast Can You React? *158*
      Is the Water Warm or Cold? *156*
   Perception
      Floating Piece of Finger *152*
      Swaying Cardboard *151*
   The Human Eye
      Are we Partially Blind? *149*
Biology-Human *149*
BODY BUILD OF MEN vs WOMEN *161*
BOTTLE ON THE BACK SWING? *113*
Boyancy Force *144*
Bromine *68*
BUOYANCY
   Melting Point
      Water Candle *52*
Buoyancy Force *51*
BURN PAPER WITH ICE *61*

C

CAN WE GET THE CORK OUT? *121*
CAN WE GET THE CORK OUT? (II) *122*
CAPILARITY *45*
Capillary Action *43*
Carbon Disulfide *62*
Carbonate Ions *58*
Carbontetrachloride *47*, *53*, *73*
Cardboard *99*
CATCH THE DOLLAR BILL *157*
CENTER OF GRAVITY
   Adhesion
      Center-Seeking Paper *103*

Air Resistance
  Stand a Dollar Bill on Your Finger *105*
Human Body
  Stuck To The Wall *102*
  The Unreachable Cup *101*
Shock Absorbers
  The Standing Matchbox *104*
Spinning Objects
  Kick A Straight Line *106*
  Stand a Raw Egg on Its Head *107*
  The Plate Carousel *100*
  Where is the Balancing Point? *99*
Center of Gravity
  98, 99, 100, 101, 102, 104, 106, 107, 131, 161
Center of Rotation, *139*
CENTRIPETAL FORCE *135, 136*
Cerebellum *159*
Cerebrum *159*
CHARACTERISTICS OF MATTER *35, 36*
  Absorption
    Funny Toothpicks *43*
  Bouyancy
    Letter Scale *51*
    Water Candle *52*
  Capillarity
    Soap Film *45*
  Colloids States of Matter
    Milk from Water and Oil *37*
  Density
    Floating Oil Sphere *46*
    Unlevel Communicating Tube *48*
    Which Will Float Here? *47*
  Density Immiscibillity
    Clearing Spot *41*
  Density of Gasses
    Floating Soap Bubble *53*
  Kindling Point
    Exact Paper Circle *42*
  Molecular Nature
    Cooling Rubber Band *35*
  Molecular Spacing
    Is The Cup Really Full? *39*
  Sinking & Floating
    Magnetic Finger *40*
  States of Matter
    Three Stages of Wood *36*
  Steam & Moisture State of Matter *38*
    Invisible Steam *38*
  Surface Tension
    How Many Pennies Can Go In *44*
  Vapor Pressure
    Drinking Bird *54*
  Water Pressure
    Communicating Cylinders *49*
    Plastic Tubing Water Level *50*
Charcoal *36*
CHEMISTRY

Combustion
  Exploding Balloons *59*
Endothermic Reaction
  Sticky Board *70*
Exothermic Reaction
  Delayed Explosion *64*
Flame and Burning
  Human Flame Thrower *66*
  Magic Candle *67*
Gas Production
  Pottasium Chloride Bombs *62*
High Polymers
  Nylon Thread Out of Two Liquids *73*
Kindling Point
  Straw Flame Thrower *65*
Oxidation Reaction
  Draw with Fire *56*
Percipitates
  Water into Wine, Milk, & Beer *58*
Physical Change
  Walk Through A Hole in Ordinary Notebook Paper *55*
Reaction Energy
  Touch A Cracker *69*
Reduction Reaction
  Red Rose into A White Rose *57*
Solubility
  Amonia Fountain *60*
Spontaneous Combustion
  Burn Paper With Ice *61*
Sponteneous Combustion
  Burn a Piece of Metal in Water *63*
Technological Application
  Sticky Matches *74*
Unsaturated Bonds
  Color Absorbing Bacon *72*
CHEMISTRY
  Reaction Energy
    Glowing Aluminum *68*
CHEMSTRY
  Reversible Reaction
    Shaking the Blues *71*
Christian Johann Doppler *96*
Chromate *74*
Circuit Breakers *87*
CIRCUITS
  Balloon Fuse *87*
Carbondioxide *53*
COHESION & ADHESION. See *CAPILARITY*
Cohesive Forces *103*
Coke from Coal *36*
COLLOIDS *37*
COMBUSTION. See *PHOTOCHEMICAL REACTION: Combustions*
  Exothermic Reaction *64*
  Kindling Point *65*
Communicating Tube *48*

Invitations to Science Inquiry – Supplement – Page 164

COMPRESSION
    Water Pressure *123*
Condensed Droplets. *38*
Cones *153*
Conservation of Angular Momentum *140*, *139*
CONSERVATION OF MOMENTUM *137*
    Test Tube Cannon *138*
Convex Meniscus *44*
Corkboard *99*
Cotton *39*

CRACK A NUT *130*
Critical Angle *92*, *94*
CRUSH THE CAN BY STANDING ON IT? *127*
CURRENT ELECTRICITY
    Circuits
        Balloon Fuse *87*
CYCLONE *34*

D

Decanedioxyl Chloride *73*
DENSITY *48*
    Bouyancy *51*
    Immiscibillity *41*
    Water Pressure *49*, *50*
    Which Will Float Here? *47*
Density *145*
DENSITY OF GASES *53*
DETERMINE THE ROCK'S VOLUME? *145*
Dextrose *71*
Dining Plate *100*
DIRECTIONAL OSCILLATIONS *129*
Dollar Bill *105*
DOPPLER EFFECT
    Hear the Doppler Effect *96*
Doppler Effect *96*
DRAW WITH FIRE *56*
Dry Distillation *36*

E

EARTH SCIENCE
    Rocks & Density
        Determine the Rock's Volume *145*
    Specific Gravity
        What is The Rock's S.G.? *144*
    Volcanic Eruption *146*
Egg *107*
ELASTICITY *115*
Elasticity of Steel Metal. *115*
ELECTRICITY
    Static Charges
        Levitation Act *86*
ELECTROMAGNETIC INDUCTION
    Dropping Race *84*
Electromotive series *160*
Ellipse *97*

Emulsifier *37*
Emulsion *37*
ENDOTHERMIC REACTION
    Sticky Board *70*
ENERGY
    Energy Transfer
        Twin Pendulum *80*
    Nuclear Sources
        Sun-Baked Potato *76*
    Potential vs Kinetic
        The Pendulum *79*
    Solar Sources
        Start a Fire With a Magnifying Glass *78*
        Test Tube Greenhouse *77*
ENERGY TRANSFER
    The Pendulum *80*
Enthalpy *139*
Exothermic *56*, *61*
EYESIGHT *150*, *152*
    Illusion *153*
        Bird in Cage *154*
        Elliptical Pendulum Swing *155*

F

FLAME AND BURNING *66*
    Magic Candle *67*
FLASH PAPER 75
FLOWING AIR
    Bernoulli's Principle
        Leaping Egg *30*
        Lift Moth Ball With Water *31*
    Flowing Fluid *31*
        Self-Priming Siphon *32*
    Resistance
        Self-Directing Cards *33*
    Wind Energy
        Leaping Egg *30*
Foam Rubber *37*
FORCES
    Adhesive Force
        Friction *108*
    Center of Gravity
        Adhesion *103*
        Air Resistance *105*

        Human Body *101*, *102*
        Shock Absorbers *104*
        Spinning Objects *106*, *107*
        The Plate Carousel *100*
        Where is the Balancing Point? *99*
    Compression Water Pressure
        Water Hammer *123*
    Elasticity
        Conservation of Momentum *115*
    Free Fall
        Center of Gravity *126*

Friction
  Get The Cork Out II *122*
  The Invisible Glue *109*
Newton's Third Law
  Get The Cork Out *121*
Perpindicular Forces
  Crush The Can By Standing on it? *127*
Pressure vs Force
  Crack a Nut *130*
Rigidity
  Pierce a Potato with A Straw? *116*
Rotational Forces
  Torques *128*
Rotational Vibrations
  Directional Oscillations *129*
Shearing
  Needle Through A Balloon *124*
Stacking Force
  Sticky Knife *120*
Strechability
  Water Tight Ziplock Bag *125*
Strength of Corrugated Materials
  Dollar Bill Bridge *118*
Strength of Waeved Material
  Loose Knife Supports *117*
Stresses in Paper
  Hold The Burning Paper *119*
The Pendulum
  Hit The Bottle on The Back Swing? *113*
  How Many Swings Can You Get? *114*
Torques/Center of Gravity
  Burning Candle Seesaw *98*
Water Pressure
  Cardboard Bottom *110*
  Squirting Water Holes *111*
  Vortex in A Liquid *112*
FREE FALL
  Center of Gravity
    Mysteriously Moving Ball *126*
Free Fall *126*
FRICTION
  Adhesive Forces
    Magic Strip of Newspaper *108*
  Get The Cork Out II *122*
  The Invisible Glue *109*
Friction *109*

G

Gas Bubbles in Liquid *37*
GAS PRODUCTION
  Percipitates *58*
GET THE EGG IN THE GLASS *133*
GET THE THREE STATES OF MATTER OUT OF WOOD! *36*
Glucose *71*
GRAVITATION *135*, *136*

Greenhouse *77*
GROWTH
  Swollen Egg *147*

H

Halogen Fumes *68*
Handkerchief *107*
HEAR THE DOPPLER EFFECT *96*
HEAT
  Conduction
    Broken Flame *81*

I

ILLUSION
  Eyesight *153*, *154*, *155*
Illusion *154*
Immiscible *41*
Index of Refraction *93*
Induction of a Current *84*
INERTIA
  Newton's First Law *133*
Innertia *132*
Insulator *86*
IS THE BALL REPELLED? *115*
IS THE CUP REALLY FULL? *39*
IS THE WATER WARM OR COLD? *156*

K

Kerosene *47*
KICK A STRAIGHT LINE *106*
KINDLING POINT *42*
  Conduction
    Broken Flame *81*
Kindling Temperature *81*
Kindling Temperature. *61*
Kinetic Energy *79*
Knives *117*

L

Lava *146*
Law of Communicating Vessels t *49*
Lazy Susan *34*, *136*
Lead Oxide *74*
LIGHT
  Reflection
    Simple Periscope *88*
  Refraction
    Spear-Fishing, Anyone? *89*
    The Magic Oil *90*
    Total Reflection *91*, *92*
  Refraction & Scattering
    Make Your Own Rainbow *94*
  Total Reflection
    Reflecting Brick Wall *93*

Liquid Pressure *111*
Lycopodium Powder *65*

M

Magma *146*
Magnesium Powder *146*
Magnesium Ribbon *146*
MAGNETIC LINES OF FORCE
   The Mysteriously Moving Needle *83*
MAGNETISM
   Electromagnetic Induction
     Dropping Race *84*
   Magnetic Lines of Force
     Mysteriously Moving Needle *83*
MAKE A FIRE CYCLONE *34*
MAKE A NYLON THREAD OUT OF TWO
     LIQUIDS *73*
MAKE MILK FROM WATER AND OIL *37*
MAKE YOUR OWN LETTER SCALE! *51*
MAKE YOUR OWN RAINBOW *94*
(HOW) MANY SWINGS CAN YOU GET? *114*
Matchbox *104*
Meniscus *48*
Mercury *47*
Methane and Hydrogen *64*
Methylene Blue *71*
Mirages *92*
Mirrors *88*
Miscible *41*
MOLECULAR NATURE *35*
MOLECULAR SPACING *39*
MOMENT OF INERTIA *139*
   The Tin Can Race *141*
Moment of Inertia *141*
Momentum *115*, *137*
MUSCLE COORDINATION *159*, *160*

N

Sodium Carbonate *58*
Sodium Peroxide *61*
Sodium Bicarbonate *58*
Needle Compass *83*
Nerve Endings *156*
NERVOUS SYSTEM
   Reaction Time
     Catch The Dollar Bill *157*
   Touch *156*
Newsprint Paper *106*
NEWTON'S FIRST LAW
   Another Balancing Act *131*
   Pull The Dollar Bill Out *132*
NEWTON'S FIRST LAW *133*
NEWTON'S LAW - FIRST PART : *123*
NEWTON'S THIRD LAW *121*, *134*
Nitrogen Tri-iodide *69*
Nylon-610 *73*

O

ORGANIC CHEMISTRY
   High Polymers
     Nylon Thread out of Two Liquids *73*
   Unsaturated Chemistry *72*
OSMOSIS *147*
Osmotic Pressure *45*
OXIDATION REACTION *56*

P

Parabolic Disc *76*
Parallax *151*
Parallelogram *103*
PERCEPTION *151*
   Eyesight *152*
Periscope *88*
PERPENDICULAR FORCES *127*
Phenolphtalein *58*
Phenolphtalein *63*
Photo-Sensitive *59*
PHOTOCHEMICAL REACTION
   Combustions *59*
PHYSICAL CHANGE *55*
PIERCE A POTATO WITH A STRAW? *116*
PITCH
   Pluck a Rubber Band *95*
Pivot *98*
Pivot Point *100*, *101*
Plane of Rotation *140*
PLANTS
   Growth
     Swollen Egg *147*
Plastic Comb *85*
PLUCK A RUBBER BAND *95*
Polyacrylate *39*
Polymerization *73*
Pop Can *26*
Potassium Chlorate *74*
Potassium Hydroxide *71*
Potassium Nitrate *56*
Potato *116*
Potential Difference *160*
Potential Energy *79*
POTENTIAL VS KINETIC
   The Pendulum
     Will the Heavy Brick Hit Your Nose? *79*
PRECESSION *140*
Precipitated *58*
PRESSURE VS FORCE *130*
PULL THE DOLLAR BILL OUT?! *132*
PUSH A NEEDLE THROUGH A BALLOON? *124*
PYREX glass *90*

R

RADIATION ABSORPTION

Which Coin Will Stay on Longer? *82*
REACTION ENERGY
   Glowing Aluminum *68*
   Touch a Cracker *69*
REACTION TIME
   Nervous System *157*, *158*
RECOIL *138*
REDUCTION REACTION *57*
REFLECTION
   Simple Periscope *88*
   The Elliptical Wonder *97*
REFRACTION
   Reflection
      Why Do We See Two Coins? *91*
   Spear-Fishing, Anyone? *89*
   Total Reflection
      Reflecting Brick Wall *93*
      Use Water as Mirror? *92*
Refraction *91*
REFRACTION & SCATTERING
   Total Reflection
      Make Your Own Rainbow *94*
Refraction Index *89*, *92*
Refractive Index *90*
Repulsion Force *86*
Response *157*
REVERSIBLE REACTION
   Shaking the Blues *71*
RIGIDITY *116*
Rods *153*
ROTATIONAL FORCES
   Confused Twirling Paper *128*
Rotational Momentum. See **Angular Momentum**
ROTATIONAL VIBRATIONS
   Yip-Yip Stick *129*
Rubber Band *95*
Rubber Cement *108*

S

Sebacyl Chloride *73*
Semi-Permeable Membrane *147*
SHEARING
   Needle Through a Balloon *124*
Shearing *124*
SHOCK ABSORBERS
   Center of Gravity *104*
SIMULATE A VOLCANO ERUPTION *146*
SINKING & FLOATING *40*
Smoke *28*
Sulfurdioxide *57*
Sodium Hydroxide *63*
Sodium Metal *63*
Sodium Silicate *74*
SOLAR SOURCES *77*
   Start a Fire With A Magnifying Glass *78*
SOLUBILITY. See **CHEMISTRY: Solubility:**
   Ammonia Fountain
SOUND
   Doppler Effect
      Hear the Doppler Effect *96*
   Pitch
      Pluck a Rubber Band *95*
   Reflection
      The Elliptical Wonder *97*
SPACE SCIENCE
   Angular Momentum
      Moment of Inertia *141*
      Spinning Football *139*
   Angular Momentum Precession
      Human Gyroscope *140*
   Centripetal Force
      Gravitation *135*, *136*
   Conservation of Momentum *137*
   Consevation of Momentum
      Recoil *138*
   Inertia
      Newton's First Law *133*
   Newton's First Law
      Another Balancing Act *131*
      Pull the Dollar Bill Out *132*
   Newton's Third Law
      Straw Rocket *134*
   Weightlessness
      The Cup of Coffee Drop *142*
      The Weightless Nail *143*
SPEAR-FISHING, ANYONE? *89*
SPECIFIC GRAVITY *144*
SPINNING OBJECTS
   Center of Gravity *107*
   Kick A Straight Line *106*
SPONTANEOUS COMBUSTION. See **CHEMISTRY: Spontaneous Combustion: Burn Paper With Ice**
   Gas Production *62*
STACKING FORCES
   Sticky Knife *120*
STAND A DOLLAR BILL ON YOUR FINGER *105*
STAND A RAW EGG ON ITS HEAD *107*
STATES OF MATTER *36*
STATIC CHARGES
   Levetation Act *86*
STATIC ELECTRICITY
   Attraction of Uncharged Objects
      Electric Meter Stick *85*
Status Nacendi *61*
STEAM & MOISTURE *38*
Stimulus *157*
STRENGTH OF CORRUGATED MATERIAL
   Dollar Bill Bridge *118*
STRENGTH OF WEAVED MATERIAL *117*
STRESSES IN PAPER *119*
STRETCHABILITY *125*
STUCK TO THE WALL? *102*

Styrofoam *37*
Sun's Energy *76*
SURFACE TENSION *44*
Suspension *37*

T

Talcum Powder *108*
Tension *119*
Tetrachloroethane *73*
THE INVISIBLE GLUE *109*
THE AMMONIA FOUNTAIN *60*
THE BALLOON FUSE *87*
THE BALLOON IN THE JAR *25*
The Blind Spot *149*
THE BURNING PAPER? *119*
THE CARDBOARD BOTTOM *110*
THE CENTER-SEEKING PAPER *103*
THE CLEARING SPOT *41*
THE COMMUNICATING CYLINDERS *49*
THE CONFUSED TWIRLING PAPER *128*
THE COOLING RUBBER BAND *35*
THE CRUSHING POP CAN *26*
THE CUP OF COFFEE DROP *142*
THE DOLLAR BILL BRIDGE *118*
THE DRINKING BIRD *54*
THE DROPPING RACE *84*
THE ELLIPTICAL PENDULUM SWING *155*
THE ELLIPTICAL WONDER *97*
THE EXACT PAPER CIRCLE *42*
THE EXPLODING BALLOONS *59*
THE FLOATING OIL SPHERE *46*
THE FLOATING SOAP BUBBLE *53*
THE FLOATING PIECE OF FINGER *152*
THE FUNNY TOOTHPICKS *43*
THE GRAVITY MACHINE (I) *135*
THE GRAVITY MACHINE (II) *136*
THE HAND IS QUICKER THAN THE EYE *153*
THE HUMAN EYE *149*
THE HUMAN FLAME THROWER *66*
THE HUMAN GYROSCOPE *140*
THE INVISIBLE STEAM *38*
THE KICKING FROG LEG *160*
THE L00SE KNIFE SUPPORTS *117*
THE LEVITATION ACT *86*
THE MAGIC STRIP OF NEWSPAPER *108*
THE MAGNETIC FINGER *40*
THE MYSTERIOUSLY MOVING BALL *126*
THE MYSTERIOUSLY MOVING NEEDLE *83*
THE OUTPOUR RACE *112*
THE PENDULUM *79*
   Hit The Bottle on The Back Swing? *113*
   How Many Swings Can You Get? *114*
THE PLASTIC TUBING WATER LEVEL *50*
THE PLATE CAROUSEL *100*
THE POTASSIUM CHLORATE BOMBS *62*
THE SELF-DIRECTING CARDS *33*

THE SIMPLE PERISCOPE *88*
THE SPINNING FOOTBALL *139*
THE SQUIRTING WATER HOLES *111*
THE STANDING MATCHBOX *104*
THE STICKY KNIFE *120*
THE STRAW FLAME THROWER *65*
THE STRAW ROCKET *134*
THE STRONG SOAP FILM *45*
THE SWAYING CARDBOARD *151*
THE SWOLLEN EGG *147*
THE TEST TUBE CANNON *138*
THE TEST TUBE GREENHOUSE *77*
THE TIN CAN RACE *141*
THE TWIN PENDULUM *80*
THE UNCONTROLLABLE FOOT *159*
THE UNLEVEL COMMUNICATING TUBE *48*
THE WATER CANDLE? *52*
THE WATER HAMMER *123*
THE WATER-TIGHT ZIPLOCK BAG *125*
THE WEIGHTLESS NAIL *143*
THE WEIGHTLESS WATER *24*
THE YIP-YIP STICK *129*
Torque or Moment *98*
TORQUES *128*
   Center of Gravity
      Burning Candle Seesaw *98*
TOTAL REFLECTION *92*
   Make Your Own Rainbow *94*
Total Reflection *91*, *94*
TOUCH
   Nervous System *156*
TRANSFERABILITY OF PRESSURE *27*
TURN A LITTLE WATER INTO A LOT OF LEMONADE *27*
TURN A RED ROSE INTO A WHITE ONE *57*

U

USE WATER AS A MIRROR? *92*

V

VAPOR PRESSURE
   Heat of Vaporization *54*
Vapor Pressure *54*
Venturi tube *31*
VOLCANIC ERUPTION *146*
Vortex *112*. See also **CYCLONE**
VORTEX IN A LIQUID
   Outpour Race *112*

W

WALK THROUGH A HOLE IN ORDINARY NOTEBOOK PAPER *55*
Water Absorbent. See **Cotton**
WATER PRESSURE *50*
   Cardboard Bottom *110*

   Squirting Water Holes *111*
   Vortex in a Liquid
      Outpour Race *112*
Water Vapour *38*
WEATHER
   Cyclone
      Fire Cyclone *34*
WEIGHTLESSNESS *143*
Wesson Oil *90*
WHAT IS THE ROCK'S S.G.? *144*
WHERE IS THE BALANCING POINT? *99*
WHICH WILL FLOAT WHERE? *47*
WILL THE HEAVY BRICK HIT YOUR NOSE? *79*

ORDER FORM 99/00

QTY	Item No	Description	Unit Price	Total
		Books		
	B1	Invitations to Science Inquiry-2nd Ed. Dr. Tik Liem	$ 52.50	
	B2	Invitations to Science Inquiry-Supplement 2nd Ed.	$ 27.50	
	B4	Turning Kids On to Science in the Home Dr. Tik Liem (set of four books)	$ 78.00	
	B4-1	Volume 1-Our Environment (186 pages)	$ 27.50	
	B4-2	Volume 2-Energy (168 pages)	$ 24.00	
	B4-3	Volume 3-Forces & Motion (126 pages)	$ 20.50	
	B4-4	Volume 4-Living Things (71 pages)	$ 17.50	
		Videos		
		Accompanies: Our Environment Book (B4-1)		
	VT-1	How to do Science Discrepant Events on AIR	$ 12.95	
	VT-2	How to do SDE on Flowing Air and Weather	$ 12.95	
	VT-3	How to do SDE on Characteristics of Matter	$ 12.95	
	VT-4	How to do SDE on Chemistry	$ 12.95	
		Accompanies: Energy Book (B4-2)		
	VT-5	How to do SDE on Energy & Heat	$ 12.95	
	VT-6	How to do SDE on Magnetism & Electricity	$ 12.95	
	VT-7	How to do SDE on Light and Sound	$ 12.95	
		Accompanies: Forces & Motion Book (B4-3)		
	VT-8	How to do SDE on Forces	$ 12.95	
	VT-9	How to do SDE on Space and Earth Science	$ 12.95	
		Accompanies: Living Things Book (B4-4)		
	VT-10	How to do SDE on Plants and Human Biology	$ 12.95	
	VT1-10	How to do Science Discrepant Events-Series of 10	$ 125.00	
		Multimedia		
	CD-1	How to do Science Discrepant Events on AIR-CD ROM (Windows 9x or Mac O.S-select one)	$ 50.00	

Please include Shipping and Handling (add 15 %) (Ca Res. Add 7.75% tax)

Ship to:
Name:_____
Address:_____
City:_____St:_____Zip:_____
Telephone#:_____

Total _____

Send checks to:
Science Inquiry
6290 Butterfield Way
Placerville, CA 95667
phone: 530-295-3338
fax: 530-295-3334
www.scienceinquiry.com

| **Send Invoice to:** | **Send Materials to:** |

_____
(Name)

_____
(Address)

_____
(City)   (State)   (Zip)

_____
(Tel)

_____
(Name)

_____
(Address)

_____
(City)   (State)   (Zip)

_____
(Tel)